Thank you Thomas, Thankful for your passion

Rob

Aug 13/19

ADVANCE PRAISE

As a former "farmer of consequence," I was often curious and at times alarmed by the public perceptions toward food. I have valued the leadership of those who have stepped up and worked hard to educate the consuming public about how our agriculture and food sector operate. Rob Saik is one of those leaders. Rob is an unrelenting seeker of knowledge and purveyor of facts. He is a gifted storyteller and through creativity and enthusiasm has parlayed his experiences to inform and educate about what matters most—our ability to sustainably feed a growing global population.

STUART CULLUM, PRESIDENT; OLDS COLLEGE

As farmers we struggle sharing the complexities and risks associated with running a modern farming operation. I think Rob does a good job of capturing life on the farm of today and I really like the way he painted a picture of where agriculture is headed. In FOOD 5.0, Rob shares that there is no "one way" and that we must adapt all kinds of science and creativity to ensure we farmers, can feed the future.

JIM PALLISTER, PALLISTER FARMS

As a farmer, operating a multigenerational farm, it is good to have someone communicate the challenges and opportunities in today's agriculture world. In FOOD 5.0, Rob does a good job of capturing where we came from and where we are heading. This is the kind of information we need people in cities to understand about the realities of food production.

JEFF CARLSON, CARLSON AGRICULTURAL ENT. LTD.

FOOD 5.0 is a great book for those involved in agriculture and even a better one for those not directly associated with farming. On our farm, we are constantly working to integrate new technology so we can farm more sustainably. That means both environmentally and economically. Rob has always been a leader in agriculture technology integration, this book is a great look into the crystal ball of agriculture to 2050.

TERRY ABERHART, ABERHART FARMS

Rob's awesome gift is that he sees agriculture from the inside, while his entrepreneurial side appreciates the perspective of the consumer from the outside. His heart (and brilliant mind) truly are in the right place—he wants a better agriculture, a better Earth, and a better humanity through better nutrition. Food 5.0 combines historical perspective to farming while painting a picture of the future of food production. With stories, simple explanations, and the occasional rant, this book has a message for everyone who eats. And last I checked, that's everyone!

DAMIAN MASON, FARMER, SPEAKER, HOST OF THE BUSINESS OF AGRICULTURE PODCAST, AND AUTHOR OF DO BUSINESS BETTER

FOOD 5.0

FOOD 5.0

How We Feed The Future

ROBERT D. SAIK

LIONCREST
PUBLISHING

FOOD 5.0

How We Feed the Future

ISBN 978-1-5445-0450-6 *Hardcover*

978-1-5445-0451-3 *Paperback*

978-1-5445-0452-0 *Ebook*

This book is dedicated to my children:

Charlene, Allison, Nicholas, and Laura

CONTENTS

GRATITUDE

Lori Rosehill, thank you for your love and support and for being patient with my endless agricultural distractions.

To my entire family, for encouraging me and sharing me with the farming community.

Mom, thanks for instilling in me a "get it done" attitude—love you!

To Shelley Myers, my long-suffering, amazing, executive assistant—I could not do what I do without you.

Jerry Stoller, you taught me more about farming than anyone.

Dan Sullivan and Peter Diamandis, thanks for providing the environment where I can dream and stretch my wings.

To the Agri-Trend family. For twenty years we worked to build something amazing. The bonds of friendship and respect will last a lifetime. To the leadership team, the support team, the programmers, the senior coaches, and all the coaches; I am so very proud of you and of the passionate legacy that lives on in you today.

Most importantly, I want to acknowledge farmers.

This book is about you. It is about where you came from and where you are going. I hope, in some small way, people who read this book will look upon the men and women who work the land with a greater level of understanding and respect. Thank you for allowing me to be a part of your world.

"The most important thing a farmer leaves on this earth when he leaves this earth is more earth."

ROBERT SAIK

INTRODUCTION

When you woke up this morning, did you worry about a plague?

Did you worry about war?

Did you worry about famine?

I'm guessing you didn't. I know I didn't; well not yet.

We're among the first generations to be blessed with such peace in our lives—which is something we should be grateful for. Looking ahead at the next thirty years, however, between now and 2050, concerns me.

As I look to the future, I'm not necessarily concerned about a plague, war, or famine, specifically; however, I am concerned about agriculture's ability to feed the future. Not whether agriculture can feed the future, but rather will agriculture be ALLOWED to feed the future.

You see, agriculture is the glue of civilizations, for without food, you have anarchy, and I see great challenges facing the sector in the near future. Never in the history of mankind has agriculture faced a challenge like we're going to face in the next thirty years: feeding our ever-growing population hopefully without the impact of war, plague, or famine.

Historically, war, plague, and famine have wiped out millions and millions of humans. However, in the last century, wars have diminished, plagues have lessened, and famine has been greatly reduced largely due to the adoption of science in agriculture.

People believe the global population will continue to rise forever, but according to UN demographers and sites like Gapminder.org, the population will level out at around eleven billion people, likely around the year 2100. Even more conservative demographers predict the population will grow to between nine and ten billion people by 2050, then level out.

The real challenge lies in feeding the population between now and 2050, while the population grows from seven billion to nine or ten billion people. Our pinch point occurs within the next thirty years.

How are we going to feed ten billion people? My belief is that agriculture can again (as in the days of Norman Borlaug[1]) rise and prove Thomas Malthus[2] wrong. I have worked in agriculture my entire life, and despite the great challenges I see in the future, I also see great opportunities.

A few years ago, in Argentina, I was interviewing a scientist who was working on isolating genetic traits that would increase drought tolerance in crops. When discussing the population rise to 2050, she said, "We have three choices. We can kill people, which we have a pretty good track record of doing, we can implement mandatory population control, or we can feed people."

I doubt that killing people is a solution; although the First World War managed to do so to the tune of thirty-six million. I hope that plagues, like the Spanish Flu that wiped out fifty million people are things of the past.

Mandatory birth control, such as the one-child policy in China, is not that palatable to most people; and it really did not work that well.

---

1   Dr. Norman Borlaug, an agronomist and humanitarian known as "The Father of the Green Revolution," won the Nobel Peace Prize in 1970 for his work on developing wheat strains that resisted drought and disease. Learn more about his foundation at www.normanborlaug.org.

2   Thomas Robert Malthus was an 18th century cleric, scholar, and author of *An Essay on the Principle of Population,* in which he wrote, "The power of population is indefinitely greater than the power in the earth to produce subsistence for man." Malthus T.R. 1798, *An Essay on the Principle of Population,* Oxford World's Classics reprint, p. 13.

So, the logical solution to sustain a population of nine to eleven billion is to figure out how to feed them. It's going to take all the tools in our toolbox, and we need to ensure that we do so in a sustainable fashion. In fact, so long as we have people on the planet, we must ensure that agriculture is *infinitely sustainable*.

Consider the numbers:

- To meet the global demand that we *will* encounter in the next thirty years, we must grow the equivalent of the amount of food we've grown in the past ten thousand years.
- By 2050, all countries will need to increase food production by 60–70 percent.

Most countries of the world, especially those that will experience the highest levels of population growth, are net importers of food. These countries need to produce more food to feed their populations, while simultaneously depending upon the exporting nations, such as Canada, Brazil, and New Zealand, to make up any gaps in supply.

When it comes to how I look at the world, I have to admit that I'm a cornucopian. I believe the Earth has enough resources to feed the people on the planet—if we use the technology available to manage them. People look to Silicon Valley for the latest developments in technology, yet many would not realize how often agriculture has been at the leading edge of technology adoption. You think self-driving cars are cool? Hell, farmers have had self-driving tractors for about two decades.

Agriculture has the ability to utilize technology to feed the world, by adopting everything from the aforementioned self-driving tractors, to sensor technology, to artificial intelligence, to bioengineering. In other words, agriculture has come a long way from a horse-drawn plow or a small red open seat tractor.

NOT YOUR GRANDPA'S FARM

Most people—especially those not involved in agriculture—have a romanticized view of farming. If asked, a soft smile crosses their face and they'll say, "I remember my grandpa's farm." Well, that image dancing around in your head is not rooted in reality.

ROB'S RANT

Can we just get rid of the image of the big red barn? Please!?

If you hear the word "farm" and think of a red barn, a round-fender pickup truck, a ruddy-faced guy or gal wearing bib overalls and a straw hat, you're watching the History Channel. Today's agriculture isn't your grandpa's farm.

The pace of change in agriculture and in society as a whole is increasing faster and faster all the time. People are increasingly separated from the farm. Agriculture's efficiency allows us to have divisions of labor. You don't have to wake up every morning to grow your own food and raise your own livestock because farmers are doing it for you.

Agricultural advances have been happening since humans decided to create stationary homesteads rather than live as nomadic hunters and gatherers. The agricultural efficiencies that have happened over the last century as technology advanced and new methods were invented have reshaped farming.

These technological efficiencies have allowed for economic efficiencies as well. As farmers earn more with technology, they, hopefully, create greater economic stability for their families and businesses. This correspondingly has led to economies of scale, meaning that over time farm operations have increased in size.

Not that long ago and still today in many developing nations, farmers basically fed themselves. As they adopted mechanization by the 1940s, one farmer fed about twenty other people. The stats say that today, one North American farmer feeds more than 165 people, which allows those other 164 people to do other things like become doctors, engineers, musicians, artists, or whatever.

But I think those numbers are wrong.

The stats say that one to two percent of the North American population are farmers and while that may be true, many of these reporting incomes as farmers are not full-time viable farming operations.

I would argue the number is much smaller if one considers which of these 2.1 million US and 0.9 million Canadian farms are what I call farms of consequence, that is, those that produce over 80 percent of our food.

Of our total Canadian and US population of 364 million, I would estimate the farmers of consequence to represent between 0.1 to 0.2 percent of the population, probably less.

That means farmers who run farms of consequence in the United States could fit into a very large college football stadium or two. In Canada, where I come from, the people who run farms of consequence could fit into a large professional hockey arena. Regardless of the country, that's not a lot of people.

Agriculture today is an extremely sophisticated business of high risk, yet still close to 97 or 98 percent of farms of consequence are family farms. Don't add a girl with pigtails and a freckle-faced boy to your bib-wearing farmer fantasy. It's a family *business*, an incorporated business of consequence that needs to operate in a sustainable manner.

Today's modern farm business has little to do with the image likely dancing in your head. Technology has changed the farm, like it has the TV repair business (oh wait, that's gone) or the general store (oops—gone too); it's not your grandpa's farm anymore.

Few people have a connection with the farm of today, and fewer understand what it takes to run a farm of consequence. While 0.2 percent of the population is in the true "business" of farming, many of the other 99.8 percent are often quick to criticize "big" ag, factory farms, and science such as GMOs, without understanding the history, science, and economics behind those modern, technologically advanced farming practices.

If you're one of those people, I'd like you to open your mind and read the pages of this book; I'm going to share some of the science and technology behind the agricultural practices that will enable us to feed the population of 2050.

Hopefully, as you read this book, you will begin to

understand what it takes to run a farm of consequence and will develop empathy for this extremely small, but vitally important segment of our population.

If you are a farmer or involved in agriculture, I hope this book takes you on a journey to the future so you are better prepared for the impacts technology will continue to make on our sector.

IS YOUR FAMILY AS DISCONNECTED AS MINE?

Unless you are royalty or landed gentry, your grandfather or great grandfather, like his father before him, is likely to have worked the land to feed his family. Over time that has changed. As we moved away from the farm, we have become disconnected from agriculture. This is the case with my family and likely yours.

On my dad's side, the Saiks, my grandfather, Gido George, escaped Ukraine between the first and second Holodomor, which were Stalin's systematic extermination of Ukrainians. His Polish wife, my baba Mary, was left behind with three children and later joined Gido in Canada where they added three more children to their family. They traveled to northeastern Alberta where they were granted a quarter section (160 acres) of land, transforming it into a working farm with their bare hands.

My mom's side, the Goriuks, were also from Ukraine.

Grandfather John and Grandma Pearl raised six children. While they moved back and forth a few times between Alberta and British Columbia, most of the income they derived was from farming. My mom spent most of her childhood working on the farm.

Marshall (my dad) married Doris (my mom) and built a farming operation near the town of Innisfree, Alberta about one mile (as the crow flies) from Baba and Gido's. This is where my sister Donna and I grew up.

I often joke that I am a "PUke" because I am about 25 percent Polish and 75 percent Ukrainian. Enough of that; now let's look at what happened to the family and their relationship to agriculture.

Both sets of my grandparents were farmers. Of the twelve children they had, six became farmers. None of those farming operations are in existence today.

Of my seventeen cousins, I am the only one involved in agriculture. I have an interest in a farming operation in Uganda and have some pure-bred cattle in Saskatchewan, and have worked my entire professional career in the agricultural sector.

My other sixteen cousins would have memories of the farm, but it has been a long time since any actually spent time on a modern farming operation.

From those cousins I have thirty-two nieces and nephews. Other than my son Nick, who operates Know Ideas Media, which produces agricultural science videos, none are involved in agriculture.

In just two generations, the family that began on the farms has now birthed musicians, doctors, engineers, builders, teachers, psychologists, and entrepreneurs, but no farmers.

In just two generations, our family has gone from an intimate knowledge of where food comes from to being quite disconnected and ignorant of modern agricultural practices.

In other words, my family may be like yours. A heritage of farming, now distant, with the descendants living in urban settings.

However, if you asked my family members about what they thought about food and farming, most would have rather strong opinions, largely influenced by the church of social media.

FOOD IS THE NEW RELIGION

Let's sit down at an imaginary restaurant table together. The waitress brings the menus, and the discussion about what to eat takes place.

The paleo diner orders the steak from cows raised in the mountains of Montana that were fed on wild grass and only drank spring water. The vegan diner chooses grilled seitan from organic wheat (yep, that's a thing) and caramelized in non-GMO maple syrup (yep, apparently that's a thing too).

What do you order? Low sodium? Gluten free? High carb? Ethically raised? Fair Trade?

Years ago, people identified with their church affiliation, their profession, the car they drove, or the designer label inside (now outside) their jeans. Today, people identify with the food they eat.

We label ourselves vegetarian, low-carb, pescatarian, farm-to-table, or omnivore. Some keep their choices quiet and go about their business, but others can be downright self-righteous about what they eat, proclaiming their diet is the diet that will save the planet. Food has become the new religion.

People get uppity when the discussion around the table turns to food. Most are quick to defend their beliefs about the right diet and the way food should be produced. Yet, regardless of the food religion you belong to—paleo, vegetarian, carnivore, or vegan, pro-GMO, or organic—we can probably all agree that as long as people are on the planet, agriculture needs to be *infinitely sustainable.*

I find when I bring up infinite sustainability, it tends to stop the preaching and arguing, and people really begin to think about the factors that would contribute to agriculture's long-term sustainability.

INFINITE SUSTAINABILITY

It is exciting to see new entrepreneurs coming up with new ways of growing food. I have listened to people explain the efficiency of insect farming.

I am sure indoor vertical farming will have a place. I can see a market for in-vitro meat, and have been fascinated by the idea of protein synthesis from the air.

For this book, my focus will continue to be broad-acre agriculture (and more so dryland, broad-acre agriculture), which refers to farmers who grow staple crops on large-scale fields for food for humans and livestock, as well as fuel and fiber. I will examine infinite sustainability through that lens.

To my way of thinking, infinitely sustainable agriculture must have five components:

- **Healthy soil.** Soil is the lifeblood of any farming operation and is the first component of infinitely sustainable agriculture. This thin layer of skin that covers the earth holds water and provides nutrients and sustenance to plants and animals. To feed the population, the soil must have enough power to sustain and grow crops. It must be alive, healthy, and live forever.
- **Abundant, clean water.** Estimates indicate that agriculture uses 70 percent of the fresh water on the planet; water is essential to plant health. We need to press forward with all technologies that would enable farmers to use water more effectively.
- **Greenhouse gas balance.** Carbon dioxide, methane, and nitrous oxide are all part of the cycle of crop and livestock production. While the topic of climate change leads to heated discussions, farmers must have the tools to deal with climate change while being recognized for the work they are doing to mitigate or remove greenhouse gases.
- **Animal welfare.** In this book, my primary area of focus is broad-acre cropping. Nonetheless, many farmers make their living from the animals they raise. They care about those animals and the quality of the life of the animal. They can't make money if the animal is suffering—period. To state that animal health is important to the viability of livestock operations is to state the obvious, and I would wish more animal rights activists would take time to talk to

farmers instead of condemning without knowing them.

- **Economic viability.** If the farm isn't economically viable, you can't have environmental sustainability nor agricultural sustainability. If farmers are going broke, rather than concentrating on sustainability, they must think about their own survival. When you're surviving rather than thriving, your focus is on meeting your own short-term needs, not the long-term needs of the global population.

ROB'S RANT

How many businesses do you know that are over one hundred years old? That would be quite an accomplishment, wouldn't it? Yet I know many, many farming operations that are called century farms, meaning that they have been in the family and passed down generation to generation for over one hundred years.

It drives me crazy when I watch anti-science, anti-modern agricultural activists preach sustainability on social media to these century farmers. When they have a business that has been sustainable for a hundred years or more, then they will have earned the right to preach. In the meantime, maybe an alternative would be to shut up and talk to the farmers about sustainability—they seem to be pretty good at it.

WHAT DO I KNOW?

As mentioned, of the twelve kids raised by my grandparents, six became farmers; however, I would argue that of the six, Dad's twelve-hundred acres of cropland and four hundred head of cattle was a farm of consequence. I say this because Dad made all his income from farming.

We raised cattle, chickens, ducks, and hogs, but we also grew grain. While Dad was a cattle guy, I was always more interested in the grain side of the equation. I studied agriculture at the University of Alberta and then began a lifelong career in agriculture.

In many ways, I was at the forefront of technological developments in agriculture. In my twenties I began to manufacture and sell liquid fertilizers. My career was most impacted by working with Jerry Stoller, a genius who taught me about soil chemistry, plant physiology, and crop nutrition. It was Jerry who taught me a new language, "the language of the plant," and that understanding of plant physiology stays with me to this day.

Over the years, I learned how to balance soils and grow crops. Micronutrients were really unheard of at the time, and I played a role in creating a new era of awareness of their use in farming in Canada and the US. This opened the doors to more business opportunities and over several years, I founded two agri-retail businesses and a fertilizer distribution company and worked with a global sulfur fertilizer manufacturer while always keeping close to my love of the soil.

My interest in agriculture coincided with an interest in computers and technology. In the mid-nineties, I moved away from the sale of crop inputs. I felt there was room in the market to create a business around consulting with farmers on agronomy, and so I founded Agri-Trend with the mission of helping farmers allocate scarce resources to produce a safe, reliable, and profitable food supply in an environmentally sustainable manner.

We developed the infrastructure to provide knowledge and wisdom to farmers through a coaching network. We did not sell crop inputs; rather, farmers paid us for our agronomic, precision agriculture, carbon credit, grain marketing, and business coaching expertise.

The Agri-Trend Coaching Network was layered on top of a data system we called the Agri-Data Solution, and it went into the cloud in 2001—back when most people believed clouds stored rainwater, not data.

We grew Agri-Trend and Agri-Data into one of the most respected consulting organizations in North America. As we looked to expand globally, we transitioned to Trimble Inc., who purchased us in 2015 and is now taking our tech global.

More recently, I've taken on the role of CEO for Dot Technology Corp. based in Regina, Saskatchewan with the mandate of developing the commercialization arm of Dot autonomous farming system, which is the largest agricultural farming robot on the planet.

I have also founded a new company, AGvisorPRO, which is the "uberization" of knowledge and wisdom helping to provide instantaneous connectivity between farmers and advisors who can provide immediate answers to their questions.

I've been fortunate to travel a good chunk of the globe, working in agriculture in places like Brazil, Argentina, Australia, Ukraine, Russia, Kazakhstan, Kenya, Nigeria, and most of North America and the EU. I am a partner in a five-thousand-acre farm in Uganda, have some cattle, and am a partner in Pergiro, Inc., which is a venture capital fund primarily investing in early-stage agricultural start-ups.

I've published more than fifty articles on agriculture and agronomy, and written one book already called *The Agriculture Manifesto*. I'm writing this book because I'm passionate that people should understand more around the science of agriculture.

No science is misunderstood more so than genetic engineering in agriculture and the vilified three letters, GMO.

In 2014, I spoke at TEDx in Red Deer, Alberta to discuss genetic engineering and address the question of whether agriculture would be allowed to feed nine billion people (I recommend finding my talk online https://www.youtube.com/watch?v=xvFD6DRnoCg.)

The highlight of my career happened on May 2, 2017, when I was invited to spend six hours with Bill Gates. We talked about how we would utilize the technology that we were implementing in broad-acre agriculture to lift the forest, the farmers, and the planet to a higher level of prosperity.

ROB'S RANT

It bothers me that many consumers are scared of food. It bothers me that unscrupulous marketers are preying on the fear of the ignorant. It bothers me that people will pay way more money for adjectives on food labels than the food in the package.

I would like to have a conversation with you about agriculture. The business and the science of the modern farm. It is my wish that I can engage you with my thoughts.

It is my hope that after reading this book you will look at food labels with a more critical eye, and call "bullshit" on those labels that are spreading fear to pull dollars out of your pocket.

LET'S HAVE A CONVERSATION

I've spent my whole life in agriculture and have seen great changes over the last four decades, taking us from systems of intensive manual labor through periods of high chemical use to today's technological agriculture, which through the utilization of sensors and data, can help us grow food more precisely and more sustainably.

I'm confident that if we can feed the population of 2050, we'll easily get to 2100, but the next thirty years are critical.

Farmers must embrace the technological advances in agriculture, and perhaps more importantly, consumers must gain a clearer understanding about how the food they eat is produced. That's what I aim to cover in this book.

This book is about the modern, broad-acre agriculture practiced by 0.2 percent of the population that helps to feed the other 99.8 percent. When I talk of broad-acre farming, I am talking of the fields you observe as you fly over the plains of the US, or the prairies of western Canada, or the massive paddocks of western Australia, or the *mato grosso* of Brazil. This is where the millions of acres produce the millions of bushels of corn, soybean, wheat, canola, rice, cotton, barley, or forage that provides food, fuel, and fiber for society.

The technology and practices explained here are key to feeding us over the next thirty years and beyond. I'm going to explain how food is produced and how this type of production will allow agriculture to feed the world's growing population.

This book is not about hobby farmers, hydroponics, or urban farming. It is about the people that I know and the families I work with, it is about the *farmers of consequence*—that very thin slice of our population that almost none of us knows.

My hope is to have a conversation about where agriculture came from, where agriculture is going, and the things that are truly important in agriculture, so we can understand each other better and come to the table to solve problems rather than disagree about our food religions.

PART ONE

THE FIVE ERAS OF AGRICULTURE

Farmers tend to get a bad rap; people think they're dumb hicks who don't know what they're doing, who pour pesticides into the earth, who have been swindled into using GMO seed by big corporations, who are out of touch with climate change, and who run simplistic and unsophisticated operations.

Nothing could be further from the truth.

Since man began farming, farmers have had to constantly make decisions based on ever-changing environments and circumstances. If there's a drought, they've got to think about watering; how much to water, where to get the water from, and how not to waste the water they have. In dryland broad-acre agriculture, there is no water except what falls from the sky, so farmers are constantly thinking about how to stretch the water they are given to grow the most per acre.

If there's too much rain, and a fungus begins to grow on the leaves and the ground is so soaked that there's risk of the roots rotting, they've got to address those problems too.

Farmers also think about the day-to-day choices around crop selection, breeding, and growing techniques. They've had to evolve as the world has changed and have often been on the forefront of technological and scientific developments. Very few industries in the world are more studied than agriculture because it's an industry that impacts everyone on the planet—we all have to eat.

If you look at the iterations of agriculture, it has evolved through four prior eras from when humans shifted from a hunter-gatherer society to today. We're moving into the fifth era, and I've seen all five in my lifetime.

Agriculture 1.0 utilized muscle to grow food; a man (usually) stood behind a plow that was pulled by oxen, or a horse, or other men (or women). That era lasted for ten thousand years.

Agriculture 2.0 happened in the 1800s when tractors

36

FOOD 5.0

began to emerge. We moved from manual food production to mechanized food production.

The shift to Agriculture 3.0, the era of chemistry, really took off between World War I and World War II. Although the use of organic pesticides such as Bordeaux mixture and copper sulfate with slaked lime goes back to the 1800s, and organic elemental sulfur dates back to 4,000 BC Mesopotamia, it was the chemistry created during the early 1900s that profoundly changed agriculture.

Agriculture 4.0 was the era of biotechnology, which includes genetic engineering and the subcategories in genetic engineering that allow plant breeders to do what they've always done—search for better characteristics in crops to deal with abiotic (weather) or biotic (pest) stress and to provide characteristics that are beneficial from those crops. I would also argue that the era of genetic engineering was the era of de-chemicalization (there is a brain stretch for you).

It should be noted that there is no 4.0 in Europe. They are still largely in Agriculture 3.0, using far more chemistry per acre to grow crops because of the EU's rejection of modern scientific techniques employed by plant breeders.

Agriculture 5.0, which we're entering today, is the result

of technological advances such as smartphones, tele-communication, data management, computing power, GPS, and robotics. Agriculture is embracing technology at a rapid pace, and we're seeing a convergence of all the iterations that came before, together with today's technological advances.

We read about the broader context of technological advances in our world, but it's coming together on the farm at a break-neck speed.

These technologies are going to influence food production as we look to 2050 and beyond.

When we talk about infinitely sustainable agriculture, we must ask:

· What does the future of food look like?
· How do we ensure all future generations are fed?

Before we look at the future of food, let's spend some time on a quick review of the past iterations of agriculture, which will serve as the foundation of our look ahead.

Chapter 1

———

AGRICULTURE 1.0: MUSCLE

I was young, maybe two or three, and in the garden with Baba. It was harvest time and there in the bright blue sky, I saw an airplane. I still remember holding a deformed potato with bulges on it imagining it was an airplane as I mimicked the engine noise through my lips.

There were no supermarkets in Innisfree. There was a grocery store, but most of the food we ate came from the farm itself. And it was constant hard work. My grandparents' and then my mom's vegetable garden, like those of their neighbors, was massive. The women and children worked in the vegetable gardens while the men worked in the fields.

My sister and I picked peas until we didn't like peas anymore. I remember digging carrots, turnips, and parsnips. Oh yeah, and those potatoes. I remember being a kid and helping to dig, bag, and haul about half an acre of potatoes from the field to the cellar. We'd haul the sacks down to the cellar in the fall and then in the spring, before they totally rotted, we would haul whatever was left back out of the cellar and feed them to the pigs.

For the first part of my childhood we didn't even have running water, which made the work of canning food really hard. Then there was the butchering of the hogs or a steer, and the chickens. I saw lots of chickens doing the headless chicken dance.

And while all this was going on in the yard, the men (and women) were out working in the fields growing the crops that generated the feed for the livestock and some grain for cash sales.

At the time, most of our farm was mechanized with tractor operations in the field, but I do remember some pretty manual work on the farm—everything from slinging fertilizer bags to shoveling grain from small wooden bins. Yep—the good ole days.

HUNTER-GATHERER TO FARMER

When man moved from hunter-gatherer to farmer about ten thousand years ago, the shift brought positive and negative aspects. Communities didn't have to pack up and move when the seasons changed or in search of the next wooly mammoth to feed and clothe the tribe. Living in more solid, protective structures meant there was less threat of being eaten by a saber-toothed tiger.

But the era of agriculture ushered in the era of heavy labor, and it wasn't pretty. Heavy amounts of human toil were required to produce food. Everything was done by hand. The soil preparation, the planting, the weeding, the harvesting, and storage were all manual tasks. The majority of the population participated in food production.

What few tools they had were primitive and, again, manual. The first plows were pulled by hand, one or two men (or women) serving as the horsepower to pull the plow through the field to turn the soil and create rows to plant the seed. Even when oxen or horses began to pull the plow, a man walked behind to steer them through the field. The muscle era was an age of extreme toil and drudgery.

Through most of Agriculture 1.0, farmers weren't equipped with the tools to fight biotic or abiotic stresses. If locusts or drought hit a small area with a dense farming population, everyone's crops were ruined and the whole community suffered.

This was also an era of child labor. Children worked in

the fields alongside their parents. Women were tasked with growing crops while men went to war or were pulled away to a large-scale construction project. The muscle era of agriculture is the era of the hunch back.

THE MUSCLE ERA HASN'T ENDED EVERYWHERE

For many people in agriculture who are employed as harvest laborers and for those, primarily women, in developing countries, this is still the reality of farming today.

In countries such as Uganda, where the average age is sixteen years, or Kenya or Nigeria where the average age is eighteen, populations are set to dramatically increase. Yet much of the agriculture that will feed that growing population still is in the era of muscle. Leadership is needed in the area of agronomics and farming education to help lift these subsistence farmers to a higher level of prosperity.

While I'm not an expert on continental African agriculture, the farmers I've met there are anxious to move themselves to a higher level of prosperity, but lack some of the education and tools to do so.

There is hope. A charity I have done some work with, A Better World, has a program whereby some of its supporters have set up micro-loans through table banking for women farmers. And it's amazing to see this kind of a program work. The women farmers raise crops like passion fruit or raise chickens. The work itself is still very manual, but with the micro-loans, some of these farmers are beginning to scale.

SUBSISTENCE TO SUSTAINABLE

People in countries like Kenya went from virtually no phones to cell phones, skipping the evolution from copper lines to cellular 3G. Why can't the same thing happen with farm technology?

Every farmer across the globe deserves to utilize the technology that is available so that they could make money from their farms.

My question regarding agriculture, is why should these developing-nation farmers have to go through all the iterations experienced by other parts of the world? Can we help them leapfrog to ensure we move from the age of muscle farming to the age of abundance agriculture? I am getting a bit ahead of myself here, but I really believe there is much we can do to leverage technology, so we can free future generations from toil to enable more young rural people to become doctors, engineers, teachers, or pursue whatever profession they desire.

How can we take what we know and help move them toward Food 5.0, replacing subsistence with sustainability and prosperity?

MOVING TOWARD THE MACHINE ERA

My earliest memories of my Gido and Baba's farm was one of muscle and toil. Although for a youngster, they are pretty great memories. While I didn't realize it at the time, the memory of our family running a threshing machine was me witnessing the transition from the muscle to the machine era.

The first step in the process was pulling the binder through the field, cutting the wheat and tying the grain into sheaves, dropping them on the ground as it went.

I remember walking with my Gido behind the binder, gathering the sheaves and stacking them by hand into *stooks,* those pyramid-shaped wheat stacks that are part of the romanticized view of farming. It wasn't that romantic for us. Walking behind the binder for days at a time was hard, tiring work. Stacking the sheaves into stooks was pragmatic—if it rained, the shape of the stook caused the rain to run off instead of pooling on the grain and ruining the crop.

After the crop was harvested, we walked beside a flat-deck wagon using pitchforks to toss the stooks onto the rack. The rack would then be taken to the threshing machine. Once again, we'd use our muscles to throw the stooks into the threshing machine.

The threshing machine was powered by a belt driven off a stationary steam engine or a flywheel off a tractor, but the work of getting the sheaves into the threshing machine and moving the grain, straw, and chaff was mostly a muscle job.

The era of the horse, wagon, and man shaped our countryside. Towns sprung up every seven to ten miles to accommodate the grain elevators that bought the grain from the farmers who delivered it on wagons. This era of farming, while romanticized in media, was a hard life. As machines were invented, farmers were quick to adopt these new efficiencies.

I am glad to have witnessed the tail end of the era of mass muscle in farming, and I am just as glad that we have machines to do the work once done by hand.

ROB'S RANT

I am always amused when I come across ideologically driven people professing that we should somehow go back in time to a simpler agriculture, where you are "one with the land," and you work with the soil with your bare hands. I know two things: that kind of agriculture will not feed a 2050 population, and I know that they have never harvested a half acre of potatoes by hand.

Chapter 2

AGRICULTURE 2.0: THE MACHINE ERA

My introduction to the machine era of farming started at about the age of twelve with my dad teaching me how to start our circa 1947 one-cylinder Field-Marshall tractor. I'm not sure why we had this relic on our farm; maybe my dad was partial to the name Marshall. Anyway, starting this gem was a real procedure.

The first step was to insert a rolled piece of blotter paper into a holder, like a cigarette. Then we screwed the holder into the front cylinder of the tractor. Dad then showed me to go to the side of the tractor and rotate the flywheel until the markings on the tractor and the flywheel aligned. We took a shotgun shell filled with gunpowder and put it into another device that was also screwed into the tractor. Dad then told me to hammer the pin on the end of the device holding the shotgun shell to ignite the shell, firing off the gunpowder and creating an explosion that turned the flywheel. The flywheel caused the piston to turn, and if everything had been properly positioned and lit, the blotter paper would ignite the kerosene at the front end of the tractor, making the motor turn over. If the stack emitted smoke rings, we knew we'd done everything right and could begin our day.

Once I had learned to start the Field-Marshall one-cylinder tractor on my own, I began to work in the field. One of my jobs was harrowing the field, which meant I had to smooth out the land after extensive cultivation and seeding. At the end of the day, I looked like I'd been working in a mine, black and dusty from head to toe with only the whites of my eyes showing. You would've thought Murray McLauchlan wrote the song "Dusty Old Farmer" for me.

Of course, the machine age began a lot earlier than my experiences as a kid.

With the industrial revolution, farming, like the rest of the world, became increasingly mechanized. John Deere's steel plow, which peeled off soil rather than scraping it, was still being pulled by horses, but new tractors and threshers were being created that had a different kind of horsepower.

And agriculture is largely about horsepower; it is about energy.

ENERGY IN, ENERGY OUT

Agriculture is about the conversion of energy in versus energy out.

How much effort goes into producing an acre of crop? Back in the era of muscle, the amount of energy required to produce an acre of corn or wheat was—and still is in many parts of the world—exorbitantly high, because farmers utilized human labor or animal muscle to toil in the soil.

The machine age replaced horses and oxen with tractors for good reason: it costs far less to convert energy from fuel such as kerosene, diesel, or gasoline into horsepower, than it does to feed, water, and stable horses and oxen, and feed and pay farmhands.

Believe it or not, people protested the tractor much like people today protest and resist GMO seed. The horse industry put up a huge resistance to the tractor. At one point, the Horse Association held rallies across the country to support horses and wrote legislation outlawing tractors. Looking back, it seems ridiculous that people would protest tractors. Throughout history people have protested innovations that later turned out to be commonly used, if not downright indispensable. Believe it or not, people protested the advent of the refrigerator, preferring ice delivery to their icebox. Today, the idea of keeping food in an icebox is ridiculous. Maybe someday, the idea of protesting GMO seed will seem ridiculous, too.

As soon as farmers realized that tractors were capable of pulling a hell of a lot more than horses could, however, those protests fell on deaf ears.

The first tractors were powered by steam engines that provided the horsepower to drive the tractor, and because they had a flywheel, they could power stationary implements such as the threshing machine.

For example, the wagon of wheat sheaves that had been gathered would be taken to the threshing machine, and the threshing machine was powered by the flywheel of a steam engine tractor. People used pitchforks to throw the sheaves in the threshing machine, and the threshing machine then threshed and separated the wheat into straw and grain. The threshing machine was stationary and powered by a steam engine tractor and the whole system was being fed by muscle.

The meeting of the muscle and machine era in the 1800s is a point of convergence. A great example of this technological convergence is when George Berry motorized a previously horse-drawn thresher invented by Patrick Bell. He created a device that cut, threshed and separated or winnowed the grain, kicked the straw out the back end, and put the grain into a hopper. That particular piece of equipment combined several operations, resulting in the nomenclature that we still use today, the "combine" (except in Australia where they call it a "header").

Today's combines are massive machines with huge horsepower capable of harvesting acres and acres in a day. The image of a combine rolling across a field and pulling in swaths of grain or straight cutting crops is one of the iconic farm images from the machine era that resonates with people.

The tractor evolution took place on my family's farm as well. The Field-Marshall was eventually retired, and we kept trading up, eventually having a large, four-wheel-drive tractor with dual tires. Today, it's common to have tractors with more than six-hundred horsepower and triangular or rubberized tracks that not only provide additional traction, but simultaneously reduce soil compaction.

Over time this farm equipment has grown in size, complexity, capability, and cost—one of those high-horsepower quadtrac tractors runs up to a million dollars. Combines set farmers back $700,000 to

$1 million. High-clearance will run you north of $600,000. A hundred-foot air drill with all-variable rate and sectional shut off will be more than $650,000. A self-propelled cotton picker is over a million. The lowest-priced item for a modern farm may be the semi-truck with trailers at just over $300,000.

It's hard to grasp that each piece of equipment is the equivalent price of a house, if not several houses depending on where you buy your real estate. And the technology inside each piece of equipment means you are not going to trust that equipment to some flunky—you need to have good operators.

GOOD HELP IS HARD TO FIND

Like any business person, farmers are always watching their budget and are looking for ways to scale, save money, and increase profits. Farmers are very cognizant of something called machinery cost per acre. And while larger equipment allowed farmers to scale to gain efficiency, they faced a new problem; they needed to find qualified operators to run this expensive modern equipment, and good help is hard to find.

Most farms are in remote, rural areas where few young people want to live and work. In the muscle era, there were small communities every seven to ten miles. As tractors and combines came into agriculture, people were released from the time-consuming physical labor. And because mechanization requires fewer people to manage larger tracts of land, many communities have disappeared as people migrated to cities.

Instead of working on the farm, sons and daughters were able to become doctors, engineers, lawyers, or whatever they wanted to be. The machine era heralded the depopulation of the farming communities that is still going on today. The UN predicts that by 2050, up to 70 percent of the population will be living in cities.

With more people being pulled to the cities, finding good operators on farms poses a real challenge. When a farmer does find a qualified operator to run his expensive equipment, he'd rather put him on a 120-foot sprayer than a sixty-foot sprayer because he wants the operation to drive as large a piece of equipment as possible to take advantage of scale. This is why the trend has been to bigger and bigger equipment. From a business standpoint, upgrading and consolidating your equipment has many benefits.

The machines are getting smarter.

If you've flown over the plains of North America and looked out the window, you've probably noticed how straight the rows are. When I was young, driving a tractor in a straight line was a skill my friends and I tried to perfect. It was a revered skill of tractor operators back in the day.

You think Tesla's technology is innovative, but we have had self-driving equipment on the farm for a couple of decades!

Most of today's farm equipment has sophisticated GPS steering capability that has evolved to sub-inch accuracy. That means the tractors steer themselves and the GPS guidance is so accurate that the next pass of an eighty-foot seed drill is down to an inch of the last pass. Isn't that amazing!

As the equipment has gotten bigger and more efficient and the creature comforts in the cab better, the role of the operator has shifted from driving to one more of monitoring the computer arrays mounted in the cab that ensure the operation is going well.

YOU SHOULDN'T ALWAYS DO SOMETHING JUST BECAUSE YOU CAN

I want to spend a bit of time talking about the relationship of equipment, specifically tillage equipment, to the soil as it pertains to soil health and infinite sustainability.

In the 1950s, 1960s, and 1970s, there weren't a lot of

ways to control weeds. A common practice was *summer fallow.* That was to fallow the land during the summer.

The field was left unplanted for a season, and during the summer farmers would till the land repeatedly with a large cultivator, or plow, or disc, keeping the land black. The theory was that the crop would grow better the following season because there wouldn't be any weeds, and the soil would retain moisture because nothing had been grown there for a season.

A lot of these thoughts and theories turned out to be incorrect. Excessive tilling does control weeds, but the tillage also destroys organic matter and fractures and compacts the soil. Plant roots have difficulty penetrating the hard pans in the soil that are created by the cultivation. And, compared to today's minimum or no-tillage practices, summer fallow didn't save much water at all.

Tillage by its nature disrupts soil and, while one can make an argument for very infrequent tillage, generally we can say that tillage has a negative effect on the structure of the soil.

SOIL STRUCTURE

Soils are comprised of sand, silt, and clay. On and in the soil, there is crop residue and decomposed crop residue which forms organic matter. The living organisms interact with the residue and break it down into organic matter that is further broken down into humus. As the organisms live and die, they bring together the sand, silt, and clay into aggregates. The aggregation of the soil lends itself to *tilth*, and the tilth of the soil determines the air spaces in the soil for oxygen as well as increases the water holding capacity of the soil. Good soil tilth encourages microbial activity in the soil.

Back in 2001 I remember calling one of my buddies, Elston Solberg, who is one of the best soil scientists I know, and I said, "Elston how much carbon is in 1 percent of organic matter?" He thought a bit, did some calculations and said, "Roughly for every percent of organic matter in a six-inch layer of soil, you will fix twelve thousand pounds of carbon per acre." Then I did some math and realized that for every 1 percent increase in organic matter, farmers would fix or sequester about twenty metric tons per acre of carbon dioxide (FYI a metric ton, also known as a tonne, is 2,204 pounds while a US ton is 2,000 pounds). That's a lot of carbon we can pull out of the atmosphere, provided we do not till and blow this carbon back into the air.

Excessive tilling breaks down the soil structure and exacerbates the microbial decomposition of soil, resulting in a breakdown of organic matter. The breakdown of organic matter releases nitrogen into the soil—which is a good thing—but at the expense of greenhouse gases, such as nitrous oxide being let loose into the atmosphere, and carbon dioxide blowing off from the organic matter as it decomposes. The reduction in organic matter, coupled with the shearing of the soil, results in mass degradation of soil quality.

The tractor meant farmers could trade muscle for machine power. It meant they could till the soil many times during the year, often two or three times in the fall then another two or three times in the spring before planting seed. Combine that excessive tilling with the harsh herbicides used in the 70s and 80s and the result was often an overworked, exhausted soil.

Over time we have learned. Today, farmers use specialized equipment to seed the crop with minimum soil disturbance. One-pass direct seeding results in no-till or minimum till enabled by modern equipment combined with GPS guidance, delivering precise products to maximize productivity while protecting the soil.

Over the last decade we have had some wet years and we have had droughts. But look up, see the sky—that's the point; you can see the sky. Why? Well, there are no soil particulates in the air blocking the sun like we see in the images of rural America in the dirty 30s. Why? Because farmers have learned how to reduce tillage. Sadly, this is not the case everywhere I have traveled.

OVER-TILLAGE CONTINUES

Some areas of the world still farm as we did in North America in the last century. In recent visits to Tuscany, Italy what I saw was a far cry from the rolling green hills, straight rows of grape vines, and shady olive orchards I was expecting.

I witnessed massive soil erosion resulting from excessive tilling. The soil structure had broken down, leaving it exposed to wind and rain erosion, which caused vast amounts of silt and clay to flow downhill into tributaries.

Over subsequent trips to Kenya, I observed a move from the age of muscle to the age of the machine, but I am not sure it is a good thing for the soil. At one time, subsistence farmers in these areas would prepare their land for planting by turning the soil over with hand-held tools. More recently I have seen one-way discs mounted on tractors. I get it: one tractor with a disc can prepare many gardens and fields very quickly, but at what cost? In subtropical and tropical climates, the tillage combined with rapid microbial decomposition results in rapid soil degradation.

We have to find a way for these farmers to leapfrog the dirty 30s and find ways to grow the crop while simultaneously conserving soil.

The amount of organic matter lost in the first part

of the twentieth century is incalculable. While the machine era dramatically reduced toil on the farm, it also meant the over-tillage of soils, something we are working to correct with today's farming systems. When we migrated to Agriculture 3.0, the chemistry era, we were able to replace excessive tillage with fertilizers and herbicides.

This is the next iteration of agriculture.

Chapter 3

AGRICULTURE 3.0: THE CHEMISTRY ERA

Your body needs protein, iron, calcium, vitamins, and minerals to be healthy. Sometimes you get an infection and need an antibiotic, or mosquitoes are out and you apply bug repellant for protection. Likewise, plants need nitrogen, phosphorous, potassium, and zinc, or boron and other microelements to grow, and at times they need defense against pests or a treatment to fight disease.

One of a farmer's key jobs is to protect his crops against disease and pests. He also has to control the weeds that will choke out a crop. And he needs to ensure the crop has adequate access to the nutrients it needs to grow in a healthy manner. Weeds, diseases, and insects represent biotic stresses.

Farmers also need to fight abiotic elements—wind, rain, drought, floods, hailstorms, and other catastrophes caused by the weather.

These biotic and abiotic forces are constantly waging war against a farmer's ability to grow a crop. As we learned in the last chapter, the advent of the machine brought with it an ability to till and with that came over-tillage as farmers used mechanical means to control weeds and speed up the release of nitrogen and other nutrients from the breakdown of the organic matter through these tillage operations.

The next era of agriculture, the era of chemistry, provided an efficiency that began to displace a lot of the mechanical means and allowed farmers to finally grow crops that weren't decimated by pests.

Chemistry in agriculture is not new. For centuries, farmers used minerals such as copper sulfate or Bordeaux mixture to fight fungus and elemental sulfur against mold. Unfortunately, the chemicals associated with agriculture today are lambasted and maligned in social media without people realizing that without crop chemistry world food production would drop overnight.

Without access to modern crop inputs, farmers would not be able to even come close to sustaining the current world population, let alone look to the challenge of feeding more than nine and a half billion people. A great example of chemistry feeding the world is fertilizer.

FERTILIZER

Before modern fertilizers, farmers had to either rely on the native fertility of the soil (which eventually declines), apply manure from animals to supplement the soil, or throw a fish in with every corn seed (hard to find enough fish in the middle of Iowa).

At the end of the 1800s, farmers were running out of ways to provide nitrogen fertilizer to the crops, and there was severe concern among the scientific community that agriculture would not be able to sustain the population unless we found an external source of nitrogen.

NITROGEN

With every breath you take and with every move you make, you need nitrogen. All living organisms—humans, animals, or plants—need nitrogen to thrive.

Take a deep breath. You just breathed in 78 percent nitrogen. The thing is, this is inert nitrogen. It does nothing to us and it does nothing for plants. For plants to take up nitrogen the nitrogen must be "fixed." These fixed forms of nitrogen (ammoniacal or nitrate N) are what the plant uses.

As mentioned, one way to add fixed nitrogen to the soil is through tillage, which breaks down the organic matter in the soil releasing the nitrogen bound in that organic matter. This process is referred to as mineralization. This excess tillage has greater negative effects than positive; excessive, repeated tillage mines the soil of nitrogen, compacts the soil, destroys the tilth, and releases harmful greenhouse gases into our atmosphere.

What if you have already "mined" a lot of the N from the soil and you don't have access to enough animal or human manure—what do you do? You go after bird poop.

HOLY SHIT!

At the end of the 1800s and beginning of the 1900s, farmers turned to bird poop to help fill the nitrogen shortage in agriculture.

Believe it or not, farmers in America and Europe began importing bird poop from the Chincha Islands off the coast of Peru. Bird poop, also known as guano, is high in ammonia and when applied to the soil, that fixed form of N can help crops grow. Farmers learned that spreading bird poop on their crops increased the fertility of the soil.

Millions of birds had populated these islands for eons,

defecating their ammonia-rich guano in massive mounds along the coast. People harvested the guano in appalling conditions, and then loaded it onto ships headed to America and Europe. Predictably, demand quickly outstripped the supply.

The nitrogen shortage led to the so-called Guano Wars between Spain and Peru. Guano harvesting provided 60 percent of Peru's annual revenue. Spain wanted a cut of the action and occupied the islands, aiming to take them back from their former colony, Peru. Multiple naval battles ensued, but Spain was unable to hold its position.

But the shit really began to hit the fan when the stocks of guano began to run out.

Severe concern again rose among the scientific community that the world's population would outstrip farmers' ability to grow crops because of a nitrogen shortage. There simply weren't enough animals nor enough guano on the planet to feed the soil that fed the world's population.

HABER-BOSCH

Chemists began stridently pursuing a fertilizer solution. During the war years, German chemists Fritz Haber and Carl Bosch developed the Haber-Bosch process that uses high temperatures and pressures together with natural gas and catalysts to convert the inert nitrogen that you breathe into fixed nitrogen in the form of ammonium nitrate or urea. Today, urea fertilizer is a granular white product with 46 percent nitrogen, meaning in a hundred pounds of fertilizer, you have forty-six pounds of nitrogen fertilizer as urea.

In his book, *The Alchemy of Air*, Dennis Hager states that 50 percent of the protein in every single human comes from nitrogen fertilizer. I would argue that Haber-Bosch is the most important invention in human history, because without it, half the protein in every human being wouldn't exist, meaning that half the people on the planet would not be here without nitrogen fertilizers.

Still today, it is the Haber-Bosch process that drives agriculture globally. However, it is not without its drawbacks.

FERTILIZERS TODAY

Most of my career has been spent working in the areas of soil chemistry, crop nutrition, and plant physiology. The key to growing healthy, profitable crops is to ensure that the nutrients in the soil are in balance, and that balance must meet the nutritional demands of the crop.

I don't care if you are conventional, organic, regenerative, agroecological, or any other farming genera; the fact is plants need certain nutrients in a certain ratio to grow. Period. Full stop. The plant needs nitrogen, phosphorus, magnesium, copper, and a host of other microelements. Our job as agronomists working with farmers is to figure out how to work with the soil and fertilizers to supply the crop with the specific food it is looking for.

The fertilizers that sustain the crops that help the farmer make a profit and feed the human population are expensive—monetarily to the farm and with a high price to the environment.

This is especially true of nitrogen. The nitrogen that we apply to the soil in the form of fertilizer is not 100 percent utilized. In many parts of the world, the nitrogen is only 50 percent (or less) effective, so the other 50 percent is released into the environment. No one wants to see this happen, especially farmers, because they've paid for the nitrogen and would just as soon see it utilized. However, many farmers simply don't have enough information as to how to meet the needs of the crop while minimizing losses to the environment.

Working collaboratively with fertilizer manufacturers, farmers, researchers, and extension people, a 4R strategy has been developed to help emphasize fertilizer efficiency. The 4Rs stand for right rate, right place, right time, and right form. By concentrating on the 4Rs we can have better discussions with farmers as to how to use fertilizer to maximize profits while minimizing environmental exposure.

One of the newer chemistries that allow us to better utilize nitrogen are slow-release technologies, so the nitrogen will be fed to the crop over time. On the front end, this nitrogen is more expensive, but there's less waste because it provides a more spoon-fed type regime to the crop.

ROB'S RANT

My hackles rise when I hear people say farmers want to "douse" their crops with chemicals or "slaughter" their land with fertilizer. Their comments make no sense because chemicals and fertilizer are expensive. Farmers have to operate on razor-thin margins, and consequently the judicious application of nutrients is important.

There are implications from agriculture on greenhouse gas balance. Nitrous oxide, the N gas released from excessive tillage and from nitrogen fertilizer, has a greenhouse effect of over 300x that of carbon dioxide, so again, anything we can do in agriculture to reduce unintended losses is a benefit for the farmer and the environment.

Speaking of gas, let's have a quick look at the impact of cropping on greenhouse gases (GHG); I'd like to give you something to think about.

GREENHOUSE GAS BALANCE

The media would have us all believe, and some have actually said that carbon dioxide is a poisonous gas! Oh really! Tell that to a plant. Carbon dioxide is the life gas to plants.

We seldom hear mention of the "fertilization effect" of carbon dioxide and its impact on plant growth. Carbon dioxide is ingested by the plants to power photosynthates, and then the plants release oxygen for us to breath.

Carbon dioxide is an essential nutrient that many crops lack during the peak growing season. For example, in the Midwest during the hot July and August months, you could vastly increase crop production simply by blowing carbon dioxide through the vegetation of the corn crops. Greenhouses pump in extra CO_2 to promote more rapid growth.

One of the ways to manage greenhouse gases, then, is to have more plants growing on the face of the planet. Farmers make a huge contribution to offsetting the carbon footprint, yet rarely receive credit for it.

In Canada, the national climate figures state that agriculture contributes sixty million metric tons (MMt) of greenhouse gases to the environment every year (this is from ALL sources: manure, soil, even enteric fermentation—cow burps). Agriculture also sequesters or stores 11 MMt carbon dioxide equivalent in the soil every year. AND—most people don't think about this—farmers store an additional 79 MMt of CO_2 every year in crops we grow, and these grains are shipped all over the world.

This means that in Canada, we are over 30 MMt to the good with respect to GHG balance—a fact not many acknowledge or even realize.

Agriculture is maligned in climate change discussions, but if we can increase organic matter, we pull greenhouse gases out of the environment and the crops we grow send their oxygen back into the environment, resulting in a net win!

Organic matter in the soil sequesters roughly twenty metric tons of carbon dioxide per acre, or approximately fifty metric tons per hectare.

Consider the cycle:

· Carbon dioxide is ingested into the plant, then turned into plant constituents such as cellulose and hemicellulose, as well as sugar and starches.

- These crops then go into the marketplace for human consumption, while the crop residue remains on the land.
- Properly managed, the residues decompose and turn into organic matter.
- Organic matter is the lifeblood of healthy soil, creating the proper tilth, or the ability to hold water.

The soil and the water is used to build the "plant factories" (plant mass), and the residue of those plant factories, when they die, becomes part of the soil structure, which thusly has pulled carbon dioxide out of the atmosphere.

We grow about fifty-five million acres of canola, wheat, and barley in western Canada and export the crops all over the world. We sequester the carbon in our soils, in the grain, and then ship that grain, for example, to China. The Chinese population benefits from eating our grain, but the cycle breaks because the waste from the Chinese human population doesn't make its way back onto the western Canadian landscape. When you grow a crop on your farm and you send that crop to an export market, you're exporting the nutrients from your soil to somewhere else. Once that grain is consumed by people of the importing nation, they pee and poop out the leftovers, but those nutrients don't make it back into our system.

The exporting nations around the world, Canada, Argentina, Brazil, Australia, perhaps Ukraine, send their nutrients to the high-density populated areas of the planet and that human waste does not make it back onto the fields from which the crop was grown. Unless you have an external source of some of these nutrients or you develop crops that somehow can synthesize these nutrients, such as nitrogen-fixing plants, then you can't possibly have a full cycle.

Farmers can till the soil, releasing the nitrogen, and satisfy the short-term needs of the crops, but sacrifice the long-term structure and sustainability of the soil. Or, they can use fertilizers to provide nutrition to the plants. Properly used, fertilizers provide sustenance to crops, but used improperly, they can release greenhouse gases into the atmosphere.

You can introduce legumes such as alfalfa, clover, and peas into your crop rotation because they have special root systems that pull inert nitrogen out of the air and fix that nitrogen in the soil. They can help supply some nutrients to the soil and crops, but the downside is that they're not 100 percent nitrogen efficient. At best, they're 50 percent efficient, so while the introduction of legumes into a crop is a great idea, it doesn't satisfy the nitrogen requirements of a farmer's long-term crop rotational needs.

The only other practical way to make up the nitrogen shortfall is with an external source of nitrogen: fertilizer.

Until synthetic biology comes up with ways for crops to fix their own nitrogen, I think we are stuck with fertilizers for a while.

"Oh wait," you say. "Let's all just go ORGANIC and not use any fertilizers." Well, time for a little "cowboy" math.

COWBOY MATH

I like the term because very often it's cowboy logic that brings complex concepts into focus with simple explanations. Consider you are going to grow two hundred bushels of corn per acre.

One bushel of corn needs 1.5 pounds per acre of nitrogen; so, you will need three hundred pounds per acre of nitrogen in total.

Over ten years that's 3,000 pounds of nitrogen required to grow the crop. (You with me so far? Good.)

Where do you get the nitrogen from?

If it's coming from the soil, you have just lost 3 percent organic matter. This means your soil has significantly degraded, and you have released a bunch of green-house gasses into the atmosphere. Do you really want to do that to your soil? Is that sustainable? No!

Add nitrogen-fixing crops into the rotation you say! OK, they're only 50 percent efficient—at most—so you are still short 1,500 pounds of nitrogen per acre. Where does that come from?

Add manure you say. OK, but where do we get all the cows, chickens, and pigs to sustain the nutrients on 400 million acres of North American soil?

Without an external source of nitrogen, there is no way to sustain that long-term crop growth. The same calculation must be computed for the other nutrients a crop requires.

The long-term Achilles Heel of organic agriculture is that its dogma will not allow the use of manufactured nitrogen fertilizer. Without added N, over the long term the soil will run out. Even organic farmers need to find a way to deal with this cowboy math.

ROB'S RANT

Several years ago, I was on a visit to a 5,000-acre organic farm. We were talking about soil testing and data management, you know, exciting stuff. We were walking through the yard and there was a pump recirculating some awful smelling concoction through the tanks near his New Holland high-clearance sprayer. SPRAYER? (Yes Dorothy, organic farmers have sprayers too.) I asked what the heck was in those tanks. He said because he was not allowed to use fertilizer, he was liquifying blood and fish meal to spray onto his organic wheat to try to increase the protein of the crop.

I asked, "What do the vegans think about that?" He said, "The vegans don't know."

The vegans want NO animal agriculture. But if you asked most vegans, I would bet they want organic food. Without access to animal manure, much of organic production would decline, creating an even bigger "yield drag" (yield deficit) behind those conventional farmers properly using fertilizers.

The regenerative movement is about crop rotation (a good thing), cover crops (hard to do when your growing season is finished October 25 and it snows on Halloween) and the integration of livestock over the land. (Cool, but again, we are being told that cows = bad—really?)

So, what should we do?

Rather than ask whether a crop is organic, regenerative, etc., consumers should ask if the crop is sustainable. For agriculture to be infinitely sustainable, the nutrients that are pulled out of the soil by the crop must be put back in. For the foreseeable future, the balanced application of fertilizers will play a key role in the future of food.

CHEMICAL PESTICIDES, CROP PROTECTION

Fertilizers have made a huge difference to crop productivity, but they could not have done this alone. Greater crop yields meant a better environment for pests. Farmers don't make money feeding pests. Pests can break a farm. The chemistry era gave farmers another critical tool, pesticides.

"Pesticides" is an umbrella term that encompasses a whole family of chemistries designed to control pests. For example, you have herbicides to control herbs or weeds, you have insecticides to fight insects, you have fungicides to fight disease, and then you might have things like nematicides to fight nematodes in the soil or miticides to fight mites.

With the advent of chemistry, farmers were better able to protect their investment. Over time, the chemistry we use has significantly improved, so we are applying less product or active ingredient per acre while having a much higher degree of specificity.

For example, many of the insecticides in the early part of the chemistry era were indiscriminate. They would kill the pest and beneficial insects at the same time. Today, nobody is interested in killing beneficial insects, so the chemistry has evolved to be very specific.

The term we like to use in agriculture to describe these products we are using is crop protection products and you will hear that often if you hang around farmers.

HERBICIDES

Weeds can be devastating for a farmer. A weed is a plant growing where you don't want it: dandelions in your lawn, thistles in the cracks of your driveway, or canola growing in a wheat field.

Weeds utilize moisture and nutrients that would be otherwise available to the crop, and they crowd out the crop, so farmers are constantly battling weeds. Weeds can stay dormant in the soil for years if not decades, and just when you think you have them controlled, they'll come back and hit you again.

In my lifetime, the chemistry began on the farm with the application of broadleaf herbicides like 2,4-D, which we still use today. It is an effective product for taking out broadleaf weeds in a cereal or grass crop. When you're growing a crop like wheat, you get paid to grow wheat, not weeds. Until the chemistry era dawned, our only defense against weeds was tillage. Herbicides were manna from heaven for the farmer.

In the early days, we applied various kinds of chemistry in fairly heavy doses. We applied different herbicides several times, depending on the weed and the stage of the crop. When we planted wheat, we applied herbicide for broadleaf weed control. If a field was invaded by wild oats, millet, or mustard, we had a different herbicide for each one, and we sprayed accordingly. At the beginning of the chemistry era, it wasn't uncommon for farmers to spray a crop with three or four herbicides throughout the growing season.

Soil-Applied Chemistry

Some of the chemistry farmers used (and still use) are soil-applied products. Through the 70s and 80s, farmers combined soil-applied herbicides such as Atrazine (for corn), Triallate (for cereals), or Trifluralin (for oilseeds) with tillage. First, the land was tilled to clear it of any plant residue, then chemicals were sprayed on the land and worked in several times to mix the product into the soil. The product was very effective for the production of oilseed crops because it would control a lot of the weeds, but in those days, we put on a lot of chemicals. We applied what was called "pounds on ground" of active ingredient. In other words, the active ingredient was measured in pounds of active ingredient per acre, and it was common for us to have products that were spread up to a pound and a half of active ingredient per acre.

Today, we have very sophisticated products that are spread in grams per acre, a fraction of the quantity used in the past. Because of the advancement of chemistry, we are able to have better crop protection while reducing our environmental footprint.

Interestingly, if we lose some of the modern chemistry we have, farmers will resort back to higher rates of older chemistry along with regression to soil-incorporated products. This would be a real setback to the environment, and I hope we don't have to go back here.

SPRAYERS—A TECHNICAL MARVEL

When you see a sprayer, are you impressed? You should be.

Sprayers are a combination of Agriculture 2.0 and 3.0, for the era of chemistry could not have evolved without the development of the sprayer.

Today's high-clearance, self-propelled sprayers are amazingly sophisticated pieces of equipment capable of spraying hundreds of acres per day with a precision we could have only dreamed about in the early days of crop dusting.

Back in the day, low-flying airplanes dusted the fields with powdered insecticide to control pests such as boll weevil on cotton. We still see GPS-guided, precision airplanes spraying crops today, but the chemistry they're using is in grams per acre and is very specific and targeted to the pest.

Today the real workhorse for delivering crop protection products is the sophisticated high-clearance sprayer. The sprayer carries between one thousand and sixteen hundred gallons of water. It might have spray booms anywhere from sixty to 140 feet wide. The farmer uses five to twenty gallons of water per acre to carry grams of active ingredient onto the crop to control weeds, insects, or disease.

When you see a high-clearance sprayer on the field, know that most of the mist you see behind the sprayer boom is WATER. Far fewer chemicals are used than in the past. It's like spreading a pop can of chemical over a football field at fifteen to twenty miles per hour. The sprayer and the water carrier get this done. Amazing!

Quack Grass and Glyphosate

One of the worst weeds we had on our family farm was

quack grass. If quack grass invaded the field, it would choke out all the crops. Quack grass is a rhizobium plant, meaning it has a network of roots from which new shoots grow, so if you tried to kill quack grass with tillage, all you did was cut up the shoots and spread the quack grass further. Weeds like quack grass are the bane of any farmer, because without its control, crop yield drops dramatically.

In the 1980s, glyphosate, commercially known as Roundup, was introduced. Glyphosate, the active ingredient in Roundup is an "indiscriminate herbicide," so it kills all the plants it touches. We sprayed Roundup on areas of the fields invaded by quack grass and then let that area sit for a season. We would sacrifice part of our crop for the year to bring the land back into production the following year.

Glyphosate is a safe, efficient product that revolutionized how we grow crops today. We're able to save soil, reduce tillage, and reduce pesticide load. We were able to clean up the land, but it meant setting aside that land for a year.

Roundup was expensive when it first came on the market, up to $50 per quart; it was an amazing product that helped us deal with something that was decimating our fields. Now the cost is about ten times less, it is off patent and available from many, many companies—in other words, we have lots of choice of suppliers.

Seems like every day there are sensational headlines screaming for the death of Roundup. This is not a good thing. As farmers, glyphosate, the active ingredient in Roundup, is one of the most effective and safest tools we have to combat weeds. A weed specialist I was interviewing in Australia said that glyphosate is a "one in a hundred-year compound." It's that good.

ROB'S RANT

Why would you kill your lawn? I shake my head when I see photos from activist sources showing people spraying Roundup on their lawn. That's ridiculous—Roundup is an indiscriminate herbicide, so if you spray it on your lawn it kills your lawn; how dumb ass is that? Glyphosate works in areas where you want to keep all vegetation down, and that was the nature of the chemistry.

Before you begin citing studies about glyphosate and cancer, hear me out. Not one government reviewing agency in the world has linked Roundup or glyphosate to cancer. Not one!

Health Canada, an organization of the Canadian government, ruled in 2015 and again in 2017 that there is no correlation between the utilization of Roundup and cancer. In 2019, Brazil came out with a study that said the same thing. Their findings aligned with every major European and international government study. These are health studies, not company studies. These are health studies, not agricultural studies. Also, and again in 2019, the US EPA reaffirmed the safety of glyphosate and ruled it as a noncarcinogen.

However, if you type "Does Roundup cause cancer?" in your favorite search engine, you'll find that the Independent Agency for Research on Cancer (IARC) says it is a probable carcinogen. They're the only agency that's made that declaration. This probable carcinogen ruling puts glyphosate in the same category as hot water, shift work, and hairdressing—yep apparently hairdressing causes cancer!

There is so much to this story.

The Murder of a Molecule

The IARC isn't a national or international body that's in charge of reviewing herbicides, yet the IARC identifies substances that could be possibly carcinogenic. It doesn't base its findings on risk-benefit; it only provides hazard ratings. A body of water is a hazard; you could drown in it. But "risk" is associated with the likelihood that you would drown. The IARC has said that there is a hazard associated with hairdressing, hot water, or glyphosate, yet they ignore the exposure levels that would say the risk is small.

The whole thing stinks for three reasons:

1. The chairman of the IARC knew of a new twelve-hundred-page study looking at fifty thousand farmers and custom applicators who were exposed to Roundup glyphosate in their course of working on the farm, and there was no correlation between Roundup use and cancer. He chose to ignore this study.

2. Chris Portier was a member of the IARC reviewing agency to review glyphosate. While he was on that agency, he took $160,000 from tort lawyers to appear as a witness against glyphosate—at the same time he was serving on the committee to review the same product. He is still taking tort lawyer gigs. *Talk about conflict of interest.*

3. David Zaruk, professor at Odisee University College, best known for his investigative work as The Risk Monger, discovered that between writing the draft and writing the final IARC monogram, ten sentences had been changed in the monogram that reversed their meaning 180 degrees. They changed the intended meaning from "it doesn't cause cancer" to "it does cause cancer." And this was reported by Reuters news agencies. No one at IARC has claimed responsibility for the changes.

You aren't alone if this is news to you; very few people are aware of this. And yet tort lawyers are using the IARC as a vehicle to lay up class action lawsuits against Monsanto, who's now owned by Bayer.

I really don't care about Monsanto or Bayer. I care about the infinite sustainability of agriculture. If we take away one of the safest weed control products ever made, what are the farmer's alternatives? Are we going to return to repeatedly spraying multiple pesticides as we did back in the chemistry era? Are we going to return to excessive tillage as we did in the muscle and machine era? What is the risk-benefit of taking glyphosate away from the farmers of the world?

INSECTICIDES

Canola seed can cost as much as $13 to $15 dollars per pound. Farmers will plant three to six pounds per acre. Add fertilizer and it is not uncommon for a farmer to have $150 to $250 per acre committed before the canola even germinates. (FYI—Potato farmers will have $1,500 to $3,000 per acre invested before the potato even sprouts.) Multiply that by five hundred, a thousand, or ten thousand acres and you get an idea of the risk.

This is the same story for all crops, whether it is corn, cotton, or cauliflower.

When the canola first emerges, it is small, fragile, and vulnerable to attack by insects. Flea beetles move in from the margins of the fields and begin to chew on the young canola leaves. At the same time, wireworms and/or cutworms can attack the seeds or the young germinating plant. So, if it was your baby, what would you do?

While farmers want to protect their crops from damaging insects, they don't want to harm beneficial or non-target insects such as bees, butterflies, ladybugs, or wasps. So, we have developed seed treatments that are more targeted than simply spraying insecticide over the entire field.

There is a lot of talk about eliminating insecticide seed treatments. What most people do not realize, is that if these tools are lost, farmers will have to resort to more indiscriminate spraying.

No easy answer here. Except perhaps a genetic one; more on that in the next section.

FUNGICIDES

Just as farmers must fight weeds and protect the crop against insect invasion, they also need to protect the crop against disease.

I am a big believer in balanced nutrition to help plants fight disease and even deal with insect pressure. However, when disease pressure is high due to wind-borne spores or favorable weather conditions for disease outbreak to occur, a farmer will look to apply fungicides to protect the crop or fight pathogens.

There are many, many diseases that attack crops. Let's just look at a couple of the more serious ones.

Fusarium is a disease that has many strains affecting many crops. If corn or wheat is infected and left untreated, the result will be the production of vomitoxin, which is a mycotoxin. A mycotoxin is a naturally occurring toxin that is poisonous in very small amounts. If corn or wheat is affected by fusarium, it would render the crop useless for livestock use.

Another mycotoxin that can occur in a wide variety of crops including peanuts, cassava, corn, and wheat is aflatoxin. This naturally occurring toxin is one of the most poisonous and most carcinogenic substances known. Aflatoxin will pass from the feed to the animal, then to eggs, meat, and milk, and chronic exposure can result in gallbladder or liver cancer.

The decision to spray or not spray a fungicide is one of the hardest decisions a farmer has to make because most of the time you are making an educated guess to spray the $20 or so per acre on the crop before there is actually a sign of infection.

As a consumer, you need to understand the consequence of risk and benefit. Yes, the farmer might use a fungicide, but that might fight the disease in the crop that can really be damaging to your health.

Some of those natural organic disease compounds can be poisonous or even deadly.

ORGANIC PESTICIDES? WTH?

Before you begin this section, please understand that I believe ALL farmers want to farm more organically. I do not know any farmers who want to spend more money on fertilizer or chemicals than they need to. Seed, fertilizer, and crop protection products are expensive and anyone writing checks for these inputs wants to decrease the amount they are spending.

Organic marketers would have you believe that organic means no chemicals and that organic is better for the environment. These "memes" inflate pricing to consumers. I do not buy into these memes; nonetheless there are many ideas in organic farming worthy of further consideration. I wish we could bring the sides together to truly seek the best of all solutions. Sadly, activism means that this is unlikely to happen.

Organic farmers face the same problems non-organic farmers face.

Have you ever thought—really thought—about how organic farmers control the weeds, fungi, and insects that invade their fields?

Do you think they're getting up in the middle of the night to hand-pluck the cabbage worm off the plant?

Do they put up a sign at the end of the row that reads, "Organic Farm, Bugs Not Welcome Here"?

No, and no.

While organic farmers employ many methods to reduce their reliance on chemistry (inter-cropping, crop rotation, use of plastic mulch covers, and other cultural techniques) there is still a need for pesticides, in organic production.

Early in the twentieth century, farmers were encouraged to use a spray made from tobacco plants as a pesticide. Since tobacco is a plant, it's technically an organic treatment. But as we all know, nicotine is not only harmful to pests, it is also harmful to mammals, including humans.

But just because the food you buy is labeled organic doesn't mean that it is pesticide free. Weeds, insects, and diseases still need to be controlled.

If the crop is organic and organic farmers lack access

to herbicides, chances are, there is plenty of soil tillage involved to control the weeds which degrades the soil, releases greenhouse gasses, and you still have the diseases and the bugs to take care of.

To be clear, organic farmers use fertilizers and pesticides, but the products they use are created from plant, mineral, and animal sources instead of chemicals. These would include organic pesticides such as Bordeaux mixture, pyrethrum, and bacillus thuringiensis.

BACILLUS THURINGIENSIS

Often referred to at Bt, bacillus thuringiensis is a bacterium that occurs naturally in the soil. It has been used since the 1950s as an **organic insecticide** because it contains protein crystals that are poisonous to those caterpillars, butterflies, and moths that attack your cabbage, broccoli, and any other crop they take a shine to.

Corn borer is an example of an insect that can be controlled with Bt, as its alkaline (high pH) gut is affected when the worm eats the Bt protein, killing the bug. We have an acidic gut, therefore we digest or denature the Bt just as we would another protein we would eat, which means it kills the worms, but we humans are not affected.

Bt is an effective insecticide, but not long lasting, so

organic farmers would have to spray their crops multiple times to provide season-long protection.

BORDEAUX MIXTURE

In the nineteenth century near Bordeaux, France, a mixture of copper sulfate and lime was sprayed on the grapevines next to the roads to keep passersby from stealing and eating the grapes. It was observed that this mixture not only kept the grazers away it also prevented mildew on the grapes, and thus an organic fungicide was discovered.

Today Bordeaux mixture is used for disease control in potatoes, apples, peaches, and grapes as well as other crops. It is a preventative fungicide so must be sprayed on before the disease appears. Because it is used as a prophylactic treatment, rates applied are quite high. It will eventually wash off the plants.

When it rains, however, copper sulfate residues from the Bordeaux mixture are washed off into the ground and chemicals will accumulate in the soil where it can be harmful to earthworms, livestock, and fish. As you learned from your high school chemistry teacher, copper is an insoluble mineral, and so after it's absorbed into the ground it eventually reaches the underground water veins and streams. It's also absorbed by any ground crops nearby.

Recently some organic farmers in France have had to abandon the use of Bordeaux mixture because the very high rates of copper were poisoning the soil, resulting in environmental damage and crop production declines. Runoff can even cause sterilization of the soil—doesn't sound sustainable to me.

PYRETHRIN

On one of my trips to Kenya, while traveling to some of the higher parts of the country, I saw fields of pyrethrum chrysanthemums. They were pretty, and they did not have any bugs attacking them. That's because the flowers in the Kenyan highlands are particularly potent in pyrethrum, the extract that yields pyrethrin, which is a powerful **organic insecticide**. Use of this pyrethrin dates back a thousand years to China where farmers would crush the flower petals, sprinkling the powder on the crop.

Pyrethrin is an indiscriminate insecticide, meaning it will kill many insects—including bees—but breaks down relatively quickly in the environment. At low doses it can act as an insect repellent, but as with all chemicals, the dose makes the poison.

Pyrethroid is the synthetic equivalent to organic pyrethrin, but in a strange twist, the organic form can be more toxic to humans than the synthetic form. This is because pyrethrum has six esters versus pyrethroid's one, meaning that the organic pyrethrin stays in the liver longer, which can lead to elevated levels in the blood, which could result in hospitalization or death. It is also highly toxic to cats, so be careful when using this organic product around little Foofoo.

A FEW OTHER ORGANIC PESTICIDES

While there are a great many compounds registered for use as organic pesticides that you can research, I will leave you with a couple more.

The use of elemental sulfur dates back to the Romans. Elemental sulfur powder is approved as an organic fungicide and is good for the control of powdery mildew. It is applied in very high rates with total applications going into the millions of pounds annually in California alone.

ROB'S RANT

Strange twist. In a previous life, I worked in the sulfur industry and helped to get certification for a 90 percent sulfur 10 percent bentonite clay compound for use by organic farmers as a fertilizer and as a powder. The organic movement frowns upon the hydrocarbon sector, but ironically, the sulfur source of this fertilizer (that is approved for use by organic farmers) comes from scrubbing the sulfur dioxide from the stack of natural gas plants. (The hypocrisy is not lost on me.)

Another organic pesticide worth looking up is Spinosad, which is an insecticide to control mites, thrips, flies, ants, and other insects. It is made from the combination of two naturally occurring soil bacterium spinosyn A and spinosyn D. **It is another organic compound highly toxic to bees.**

Bottom line, organic does not mean no chemicals. And many organic pesticides are toxins, neurotoxins, and lethal toxins—and some of these potentially deadly toxins are 100 percent natural, such as mycotoxins.

MYCOTOXINS

I have already commented on mycotoxins, but I believe this is worth mentioning again because mycotoxins are the number one cause of liver cancer in the world. Mycotoxins are natural fungi and are capable of causing disease and death in both humans and other animals. They grow on crops that are weak in nutrition or don't have disease suppression. Mycotoxin invades a crop and shows up as a deformity or as a coating on the grain.

Estimates say up to two billion people or more are exposed daily to mycotoxin. People worry about parts per billion of glyphosate in their food, but they don't worry about **organic mycotoxins**.

What's even more alarming is that the organic industry doesn't have the tools to fight fungus infection the way that conventional farmers do. Invariably, your myco-toxin levels would be higher in organic food than they would be in conventional food. But the Environmental Working Group sponsored by the organic food industry will likely never run mycotoxin testing.

If you don't believe me, do your research and become an informed consumer.

SO ORGANIC MAY NOT MEAN WHAT YOU THINK

In closing this section, I want you to note a shift in some of the organic labeling. Maybe you have been seeing the words "no synthetic pesticides" on packaging.

Why?

Because many organic producers do use pesticides; they are just *organic pesticides*, but just because a product is "natural" doesn't mean it's harmless. Most organic pesticides are more toxic than Roundup, but nobody talks about that. Chemistry, including organic chemistry, used wisely, protects us from crop loss and natural pests.

ROB'S RANT

A couple years ago, I spoke at a vegetable conference in Austin, Texas. The room was full. I was nervous because I was going to talk about GMOs and this whole organic thing. Turns out the reception to my talk was very good. Afterward I met a fellow named Robert (great name) who worked as a border control officer. He asked me, "If a shipment of organic produce shows up at the border and we detect something like whiteflies in the trailer, and we have to fumigate the trailer with a gas to kill the flies, is it still organic?" Then he winked at me and said, "Happens all the time."

Don't get me wrong, I am not saying you should stop paying 30 to 600 percent more for your organic products. If it makes you feel good, do it. I won't. Until somebody shows me differently, organic farming is not a sustainable long-term agriculture practice—especially for broad-acre agriculture.

We need chemistries, organic and synthetic, that provide specific weed, insect, or disease control. We also need a way to sustain the soil with the nutrients and life it needs to produce healthy crops while reducing the environmental footprint.

To do this, we need to adopt sciences that allow us to reduce those inputs by relying on new technologies that will fight pests while increasing farm viability.

Chemistry was a solution, and it used to be the only solution we had until Agriculture 4.0 happened.

Chapter 4

AGRICULTURE 4.0: GENETIC ENGINEERING

In the age of social media and blogs, fake news and misinformation spreads faster than gossip in a small town and is just as—if not more—damaging to the community. Perhaps the biggest fumbling boondoggle in agriculture's history has been the miscommunication of genetic engineering and its tremendous contribution to agriculture. Genetic engineering is one of the most important tools that we have in our toolbox to ensure that agriculture can feed the population of 2050 and is infinitely sustainable.

Jason Lusk, professor at Purdue University, surveyed one thousand Americans in 2015 about food labels. Eighty-two percent said that if the food was GMO it should be labeled. In the same survey, 80 percent of the people said that if the food contained DNA, it should be labeled—all living material contains DNA!

The whole conversation around biotechnology and genetic engineering in particular has been hijacked by people perpetuating FUD—fear, uncertainty, and doubt. The public doesn't understand much about the science, but is deathly afraid of those three letters: GMO.

ROB'S RANT

When I am asked about my opinion regarding GMO, I say, "The only reason you would be against genetic engineering would be lack of understanding, for if you truly understood the science you could not be against GMO."

DE-CHEMICALIZATION?

Contrary to what the activists scream, the biotech era is the era of LESS chemical load in the environment; I call it the age of de-chemicalization. Genetic engineering allows us to use fewer chemicals, which you'd think the world at large would embrace with open arms, but many countries, and Europe as a whole, resist using genetically modified seed for human consumption. (You'll learn why at the end of this chapter.)

Paradoxically, they accept genetically engineered corn

and soybeans grown in North and South America to support their livestock industry, which would crash without genetically engineered crops, while not allowing their farmers to use the technology that they are feeding to their livestock—whack!

COULD THE FUTURE BE GMO—GENETICALLY MODIFIED ORGANIC?

Now, before you throw this book against the wall, please consider this. No farmer wants to spend more money than they must on fertilizer. No farmer wants to spend more money or use more pesticides than necessary to grow their crop. It's neither in their economic interest nor in the interest of the environment. The best science we have, that would allow farmers to use less fertilizer and less chemistry, is genetic engineering.

We could engineer plants to extract higher levels of nutrition from the soil, or we can engineer plants to fix their own nutrition, such as nitrogen-fixing cereal grains. We have engineered plants to fight insects and diseases, so we wouldn't have to use pesticides. In other words, genetic engineering would allow us to farm more organically, so maybe the future of agriculture is GMO, *genetically modified organic* farming.

But GMO is the devil. It's not natural. What the heck is natural in the world of breeding? Is your pug dog natural, or your cat? Everything has been bred to suit our purpose.

CROSS-POLLINATION, HYBRIDS, AND MUTAGENESIS

The origins of biotechnology and plant breeding date back to the 1800s and Gregor Mendel, a biologist, mathematician, and Augustine monk who learned about genetics by observing and cross-breeding different varieties of peas.

Through the course of time, we learned that there's a genetic code embedded inside every living being.

Chromosomes, genes, DNA, and RNA work together to make plants and animals what they are. As long as there's been agriculture, people have been breeding in a variety of ways to improve crops and animals.

OPEN POLLINATION

Open pollination is a breeding method involving the free and open flow of pollen between plants. The pollen moved by wind, birds, bees, or humans fertilizes itself or neighboring plants. Over time this randomized crossing can result in changes in the subsequent generations as the plants adapt to changes in their environment.

Farmers refer to these crops as OP varieties. Crops like

wheat and barley are examples of OP crops, as their seed can be saved and reused. However, over time this bin run or saved seed will tend to degrade in vigor, so many farmers opt to purchase "registered" (from foundation stock) or certified (genetically pure) seed from registered growers who specialize in breeding the newest and most vigorous varieties.

HYBRIDIZATION

Cross a female horse with a male donkey and you get a mule. A mule is a hybrid. You should read up on the characteristics of a mule; it's pretty nifty. Mules are very resilient and strong, but they cannot breed; such is the nature of a hybrid.

Hybrids are the specific crosses between two plants where the pollen from one variety only fertilizes another variety. The breeding of hybrid crops is largely a human activity. We do this because the vigor of hybrid corn, for example, is much better than the vigor of either parent.

Like mules, hybrid seed cannot be resown, because you get too much genetic back draft resulting in poorer crops.

So, NO, neither Monsanto, nor any seed company FORCE FARMERS to buy new seeds. Farmers buy new seeds because they want to take advantage of hybrid vigor. Or they want the newest genetics. Farmers could and do save seed for reuse; however, eventually the genetics need to be refreshed with newer, better-yielding varieties.

ROB'S RANT

If you see ANYONE on social media screaming that companies are forcing farmers to buy their seed, you will know not to believe a word they say because they don't even have a basic understanding of hybrid seeds. Time to move on from that social media conversation for sure.

MUTAGENESIS

Another process that's little known outside agricultural circles is mutagenesis. Mutagenesis creates new crops by exposing seeds or plants to something that would mutate the genetics of the plant. Mutations happen over time in nature by plants being exposed to ultra-violet light or naturally occurring radiation. In the lab, popular mutagenic breeding processes would include subjection of seeds to nuclear radiation or chemical exposure, which scrambles the genetic material.

When I was a kid, our grapefruit was mostly white-fleshed and very sour. Today, however, virtually all grapefruit is bright red and sweet. How did the grapefruit go from white to red? Through exposure to gamma nuclear radiation or irradiation! The radiation scrambled the chromosomal complex of the grapefruit seeds and poof, one of the scramblings resulted in a red and sweet-fleshed grapefruit—the Rio Red. Virtually all grapefruit that you buy in the store today owes its red flesh to mutagenesis.

Much of the barley hops used to ferment beer owes its origin to mutagenesis, and hundreds of the vegetables that you eat today owe their origin to the random scrambling through mutagenesis.

Ironically, the organic community accepts the practice of mutagenesis—the random scrambling of the chromosomal complex—but refuses to endorse the use of genetic engineering science.

I pulled this straight from Wikipedia because it says something pretty important and true:

"Several organic food and seed companies promote and sell certified organic products that were developed using both chemical and nuclear mutagenesis. Several certified organic brands, whose companies support strict labeling or outright bans on GMO-crops, market their use of branded wheat and other varietal strains which were derived from mutagenic processes without any reference to this genetic manipulation. These organic products range from mutagenic barley and wheat ingredients used in organic beers to mutagenic varieties of grapefruits sold directly to consumers as organic."

ROB'S RANT

So, here's the kicker. The organic industry is fine with the random scrambling of thousands of chromosomes by exposing seeds in a lab to nuclear radiation or carcinogenic chemicals, but is against the most modern of breeding techniques associated with genetic engineering? Riddle me that, Batman!

GENERALLY MISUNDERSTOOD ORGANISM (GMO)

Genetic engineering is nothing more and nothing less than the advancement of the breeding process.

GMO is not an ingredient.

It's a very poor descriptor for a variety of breeding techniques like transgenics, cisgenics, gene silencing, or RNAi (please, dear reader, research these) that have been made available because of the power of the computer. It is the computer that has allowed us to engineer plants. The computer has allowed us to sequence genomes, which, in turn, allows us to isolate specific genes that would release or catalyze certain enzymes or cause beneficial changes in the plant.

The first real commercial GMO success is the papaya that was genetically engineered to have resistance to the ringspot virus. This saved the Hawaiian papaya farmers. No Monsanto. No chemicals. No big companies—just farmers and scientists using technology to solve a problem.

My journey with genetics began in the mid-90s when we first were exposed to hybrid canola. Up to that point, farmers had saved and reused their seed, but yields were not really going anywhere. This hybrid stuff really kicked ass. It was more expensive, and farmers had to buy the seed every year, but once they saw the results, they were in!

Remember: hybrids have nothing to do with GMO—they are seeds from a breeding technique.

It was around 1995 when I was first introduced to a breathtaking new technology called herbicide tolerant (HT) canola. Scientists were able to create canola that was resistant to glyphosate. WHAT!?

Glyphosate (Roundup), as I explained earlier, is an indiscriminate herbicide—it kills all the plants it touches. However, scientists were able to utilize genetic engineering to create canola that was resistant to the chemical activity of glyphosate. You could therefore spray glyphosate on your canola crops, killing all the weeds and not the canola, creating a very, very clean crop. Combining HT genetically engineered technology with advanced hybrid canola sent yields upwards.

The benefits from this were staggering. We could stop spraying trifluralin on the soil. We could stop tilling the fields between three and five times prior to planting the crop. The introduction of herbicide-tolerant crops paved the way for the movement toward minimum and zero tillage. Herbicide-tolerant crops allowed a dramatic reduction of the amount of pesticide use per acre.

Every genetic trait is referred to as an event. Scientists used the event that conferred glyphosate resistance to plants meaning that farmers had a very safe and effective way of controlling weeds that decreased costs and is better for the environment.

This was huge. Biotechnology enabled the reduction of total pesticide load on the farm.

BT TECHNOLOGY

The true, important benefit of the GMO story continues to go untold: GMO crops reduce the massive amounts of insecticide previously used to have a successful yield.

You have already read about the organic insecticide called Bacillus thuringiensis or Bt. The worms eat the bacteria when they nibble on the plant, but their alkaline guts can't digest Bt, so they die.

Scientists observed this and determined that by taking certain proteins out of Bacillus thuringiensis and inserting those proteins into crops such as corn, soybeans, cotton, eggplant, cowpeas, and squash, through a genetically engineered process called transgenics, they were able to create resistance to the worms that often destroy fields of crops.

This means that crops that have been conferred with the Bt event, like Bt cotton, soybeans, corn, squash, eggplant, or cowpeas, have been genetically engineered to create insect resistance in crops, so farmers don't have to use insecticides. Our human guts use acid to break down food, so we have no trouble deconstructing the Bt protein as we do any other protein we eat.

Wouldn't you rather eat an innocuous protein than vegetables treated with insecticides?

The biggest benefit to Bt technology is found in places like Bangladesh. Bangladeshi farmers had to spray their brinjal or eggplants up to four to six times *a week* with insecticides, and most of the spraying was done by women and children. By planting Bt eggplant, they've reduced spraying by 96 percent, which means a huge reduction in insecticide exposure to humans. It's a great success story, but sadly, the general population knows nothing about it.

ROB'S RANT

I don't mean to beat a dead horse, but I have to repeat this: it's the introduction of herbicide-tolerant and insect-resistant crops that allows farmers to reduce the use of pesticide load on the land.

It is through genetic engineering and Bt technology that have thus turned farms more organic, not less.

HALTING PROGRESS

If we could utilize more biotechnology in agriculture, we would find even more ways to engineer crops that are insect resistant, drought tolerant, that are better able to fix their own nitrogen fertilizer or extract nutrients better out of the soil.

In Mexico, farmers noticed that certain corn plants produced a mucus-like substance around the root systems and that variety of corn was able to fix its own nitrogen—we'll talk more about this in chapter 6. If scientists discovered a way to put that naturally occurring phenomena into corn plants, it would lead to using less fertilizer. This is a current and on-going opportunity.

The anti-GMO movement really is a movement against perceived big ag, but what if they are wrong?

The truth is a large segment of breeding research has nothing to do with corporations.

CRIMES AGAINST HUMANITY

Scientists in Uganda and Kenya, for example, are working on diseases that are attacking their staple crops, cassava and banana. Cassava is a tuber crop that gets attacked by the brown streak mosaic virus. When you get brown streak mosaic in your cassava crop, and you grind that crop up and feed your children, the children get sick. To combat this problem, scientists in Uganda and Kenya have genetically engineered a brown-streak-resistant cassava. Unfortunately, this staple crop that is so critical during times of famine—developed by Ugandan scientists for the Ugandan people—sits behind concrete walls and barbed wire unavailable to the public. Why? Because anti-GMO organizations and activists have created FUD in the minds of politicians.

At our farm in Uganda, we have tried to grow corn without success; even when we spray the crop with insecticides, the corn simply falls over due to corn borer damage. Bt corn seed would solve the problem of the corn borer worm, but we can't plant Bt seed in Uganda because it's considered a GMO technology—and GMO seeds have not made their way into Uganda; so, we cannot grow corn for the people.

Yet another example would be bacterial wilt and black sigatoka in bananas. When I was last in Kampala, people were cutting down their banana trees because they were infected, oozing with pus. Farmers who would normally grow bananas were importing bananas from areas that weren't as heavily infected. A cure exists: a genetically engineered banana that resists bacterial wilt and black sigatoka—but it also sits behind concrete walls and barbed wire, unavailable to the public. Why? Because activists have created fear. This is immoral and wrong. Science and solutions are being withheld from people who desperately need them because other people have a preconceived notion of something they don't understand.

Genetically engineering plants to introduce higher levels of nutrients such as zinc or iron into crops, known as biofortification, holds tremendous promise. Let's look at golden rice, for example. Greenpeace activists have campaigned against golden rice since it was first developed in the late 90s.

Golden rice is fortified with beta carotene, which provides vitamin A, a nutrient that's critical for eyesight. Millions of people have died or gone blind from lack of vitamin A in their diet—tragedies that could have been alleviated with the introduction of golden rice. Since golden rice was first developed, researchers have been fighting for its release into the population. Only in 2019 have some countries like Bangladesh begun growing golden rice.

The scientist I interviewed in Argentina who spoke about the rising population is working on developing drought tolerance in crops. She found a genetic trait that conferred drought tolerance in crops, and her discovery has moved from Argentina to be developed by the Arcadia Corporation in the United States.

Most recently, a group was able to manipulate one gene in wheat to increase iron levels, and an organization called 2blades has been working to develop transgenic solutions to rust attacking wheat, which is a wicked problem.

All over the world, scientists continue to be excited about genetic engineering, yet the activists rule social media and the airwaves. Why?

WHY IS THE EU ANTI-GMO?

In the 1980s and 1990s, Monsanto made the strategic decision to shift from producing agricultural chemicals to becoming a bioscience company. Their decision paid off in spades. Their competitors, who were European companies such as BASF, Syngenta, and Bayer, continued to sell chemicals to farmers to control pests, weeds, and disease.

With the advent of Roundup and GMO seed, however, more and more farmers became Monsanto clients. The farmers were far happier controlling the weeds with a pop can of Roundup per acre than spraying buckets of chemicals several times a year on their crops. Monsanto products saved the farmers time and money, but naturally took away a lot of business from BASF, Syngenta, and Bayer.

You can imagine the backlash in the boardrooms of the EU companies, and consequently in the government seats of the European Union countries. Monsanto was quickly kicked out of the EU like an orphan child, because none of the other companies selling agricultural chemicals had made the investment in biotech that Monsanto did.

It explains why most of the major ag corporations were only too happy to let Monsanto take the "devil" pole position.

In 2018, Bayer bought Monsanto. It will be interesting to see the changes that happen in countries where Bayer has a strong foothold. Will attitudes change or will the legacy image of Monsanto be too hard to break?

There's a second reason why the EU is adamantly anti-GMO. Activist organizations such as Friends of the Earth and Greenpeace have hijacked the agricultural discussions. They play on people's emotions rather than provide factual science.

To top it off, EU regulators have decided they will design their regulations around an idea called "the precautionary principle." The precautionary principle basically says that you can't sell something or use something until you can prove it's safe.

At first blush that sounds rational, but in science you can't prove something is safe. You could give me a thousand studies that say whatever you're producing is safe, but I could say, "It's not a big enough study. It's not enough data." It will never be enough data for people who don't like it.

The precautionary principle has handcuffed the scientific community in the EU and many other countries because risk-benefits is no longer in the equation. You have to prove that something is safe, and it's just about impossible to do that.

This philosophy stemmed from a conference in 2000 that led to the Cartagena Protocol on Biosafety which basically said, "We'll govern scientific advancements under the philosophy of the precautionary principle."

The United States and interestingly, Russia, never did sign the document. While Canada and Australia were signatories, they never followed through with implementation of the plan.

I have spoken with agricultural scientists in Africa who have said the EU sponsored the environmental ministers, not the agriculture ministers, from African countries to attend the conference. At the end of the conference, each represented nation was asked to sign the Cartagena protocol. The African nations followed the lead of the EU countries (who'd paid their way) and signed the protocol. When the environmental ministers returned to their African homelands, many of the agricultural ministers said, "What have you done? You've just choked our ability to utilize science to help farmers in our African nations."

We're seeing the results of that bio summit where elitist activist groups use that as pressure to prevent nations from moving forward with genetic engineering technology that could help their people.

However, there are nations that are beginning to realize that the EU and many activists do not have the best interests of their people at heart.

Some courageous African nations are starting to flex their muscles and say, "We're going to move ahead with some science here."

Nigeria is leading the way for African countries who want to move into a new era of agricultural science.

During a visit to Nigeria I was privileged to spend time with Chief Audu, now Nigeria's Minister of Agriculture. Nigeria has agreed to release Bt cowpeas, which will be insect resistant. They're also bringing in Bt cotton, which is resistant to boll weevil.

People travel to Europe and fawn over the rolling hills and quaint villages. The truth is EU farmers don't have access to GMO crops, so they have no choice but to rely on old technology. They don't have access to herbicide-tolerant crops, so they have to till. And they spray way more insecticide than we do in Canada and the United States. The European load of active ingredient per hectare or per acre is higher than it is in North America. (Don't believe me, research environmental impact of pesticide per hectare or acre in EU versus North America; you will be surprised to find their load is higher.)

Paradoxically, while they can't grow GMO crops, they still import virtually all of their feed for the livestock industry from North and South America in the form of corn and soybeans grown with GMO technology. They know that without the American feed, the livestock sector in Europe would crumble.

EUROPE, THE MUSEUM OF AGRICULTURE?

I would not be writing Food 5.0 in Europe, because Europe doesn't have five iterations of agriculture.

The Europeans, by virtue of the NGO and activist lobbies, have grounded themselves into a post-modern agricultural breeding scenario that will prevent them from enjoying the benefits of the fourth era.

EU agricultural scientists will not be able to use any of the newest crop breeding technologies because all, including the very latest CRISPR techniques that enable precision edits to DNA, have been deemed "GMO."

Prohibited from using the new technologies that create sustainable agriculture, European farms are being trapped in the past. Many feel they are becoming the Museum of Agriculture.

HARVEST HEARTBREAK

I want to close this section on genetics with some comments and observations regarding the plight of small landholders as they work to move from subsistence to a higher level of prosperity.

What is needed is an understanding of basic agronomy coupled with the newest technology.

Take seeds for example.

One of the particularly heartbreaking things that I've witnessed is seed fraud. Farmers save their money to buy seed for their crops, and they unknowingly buy seed that's heavily diseased or genetically poor, despite its fancy packaging. Seed fraud condemns people to continued poverty. The farmers and their families go through all the effort of working in the field with their hands only to see their crops fail because the seeds aren't of the quality represented on the package.

Activists scream about farm-saved seed, but when your farm-saved seed is crap and the seed you are being sold is crap, it's hard to win.

In many areas, if farmers had access to top genetics from reputable firms, not only would yields dramatically increase, the environment would also be better off.

Today, no one should be deprived of the tools of agriculture that would allow them to farm more efficiently. To my dismay, I've seen some people in the developed world of North America and the European Union believe the best thing for small landholder farmers is to continue using the methods of the muscle era food production. Deeply rooted in colonialism, elitism continues to be prevalent in political policy that tries to keep Africa down by not promoting modern farming practices. Condemning people to the era of muscle food production is only good for those that don't have to do it.

If we are going to feed the future and ensure infinite sustainability, agriculture must learn from the past and embrace the technologies of the future. It is happening today on farms everywhere as we enter the era of technology convergence on the farm.

Chapter 5

AGRICULTURE 5.0: CONVERGENCE

I worked all summer between my junior and senior year at the University of Alberta. I saved enough money to buy a 128k IBM computer with a monochrome screen and nine-pin printer. I came home from university in the fall and set up my computer on the kitchen table of the farm.

"What's that?" my dad asked.

"That's a computer, Dad," I said, enthusiastic and proud of my purchase. "It's going to change the way we farm."

"Really? What does it do?"

"Well, it'll do spreadsheets. It'll help us to do the planning. It can do word processing."

"Huh," Dad said, unconvinced. "I don't know what that stuff is, but how much did that machine cost?"

When I told him $6,300, swear words filled the room. Dad turned around, slammed the door, and didn't talk to me for several days.

A few days later, I was out working in the field and broke an axle on one of the pieces of farm equipment. Because I wasn't a good welder, and my dad was, I brought the axle to the yard. When I showed Dad the broken part and asked him if he could fix it, he looked at me and said, "Put it in the computer."

Looking back, I can understand why my father was so upset. That $6,300 would have paid for the swather we needed to cut the crops that fall.

Despite my father's resistance to my first computer, for me it was the beginning of a lifelong interest in how we can leverage technology to make differences on the farm.

I have always been interested in following what is going on in the broader tech community, endeavoring to determine how technology would impact the farm. Through my story, you can see how technology and agriculture are converging.

From my first Palm Pilot, through to the sale of Agri-

Trend/Agri-Data to Trimble, to the launching of our newest ag technology companies Dot Farm Solutions (autonomous robotics) and AGvisorPRO, I have been surrounded with very bright people who can turn vision and ideas into technology that can help the agricultural sector.

WHEN MOORE'S LAW HIT ME IN THE HEAD!

Moore's Law states that computing capacity doubles every twelve to eighteen months, and the cost reduces by half in the same time period. I've been exposed to Moore's Law my entire working career. I have more power at my fingertips every six months. This has led to an annual ritual of buying a new piece of computing equipment or technology; however, it was my experience at Singularity University where I experienced the full force of Moore's Law.

Every ninety days, since 1994, I have flown to Toronto to attend an entrepreneurial leadership program called Strategic Coach led by Dan Sullivan. Every ninety days I review my life, my business, and my goals, and Dan leads us through thought exercises. It's a day every quarter where I slow down and think about "what if..."

In one early Strategic Coach session, I remember literally tracing my Palm Pilot on a piece of paper imagining all the "farming" things that could go in and out of the device—of course at the time, it was just a dream. But, over time that dream became the reality we are living today.

In 2012, through my involvement with Strategic Coach, I was offered the opportunity to attend a special session in Mountain View, California at Moffett Airfield. Mountain View is home to a NASA Research facility, Carnegie Mellon University Silicon Valley, Google, and most famously, Singularity University.

Singularity University is a global community of entrepreneurs, corporations, global non-profits, governments, investors, and academic institutions in more than 127 countries who explore how to address some of humanity's biggest challenges using exponential technology.

Fifty-two of us attended an intimate two-day session led by Singularity founders Peter Diamandis and Ray Kurzweil. I immediately liked Peter; he loves space, Star Trek, and we share a birthday. He is the author of the book *Abundance*—a must read if you ask me. And Ray, author of the book *How to Create a Mind*, stretched my brain like no one ever has. Together, during the two-day session, these two alternated presentations on a wide variety of technology topics such as artificial intelligence, robotics, and digital biology. My brain has never been the same.

After the first intense day, I went to my hotel and did not sleep a wink. My mind was racing. The next day was the same—so much was happening so fast. In two days, I took eighty-three pages of notes, and that's when I realized the impact Moore's Law and exponential technology would have on agriculture.

Since that session I have been a member of Peter's Abundance 360 event. Every January, Peter brings onto the stage leaders in various technology fields. In 2019 Dr. Kai-Fu Lee, author of *AI Super Powers* (another book I recommend), again rocked my world with a vision of the future that is exciting and scary at the same time.

Since that 2012 session, my passion has been to leverage advancements and developments in technology hubs such as Silicon Valley and bring it home to ag.

Peter Diamandis and Ray Kurzweil predict that by 2023—maybe sooner—a thousand-dollar laptop will calculate at the same speed as the human brain, ten to the sixteenth calculations per second. As quantum computing comes into the mix, the acceleration will increase perpetuating Moore's Law. We may be close to the limit of the number of transistors that fit on a computer chip, but quantum computing blows that out of the water. What wicked agricultural problems could be tackled with quantum computing?

At one of the Abundance 360 conferences, I listened to Chad Rigetti talk about how his company is trying to crack quantum computing. He said he would employ quantum computing to decipher the Haber-Bosch process that converts inert nitrogen to fixed nitrogen. I was shocked. Here was a computer guy speaking my language. I asked him why he used that specific example. He said, "I'm a farm kid from Moose Jaw, Saskatchewan, and the Haber-Bosch process has always interested me."

And people think farmers are dumb! Ha!

ROB'S RANT

What if we could bypass everything? Instead of turning inert nitrogen from the air into fixed nitrogen fertilizers, then applying the fertilizer to the field, then growing the crop, then feeding that crop to an animal or eating that crop for the protein, what if we instead used quantum computing to figure out how to make protein out of thin air by taking inert nitrogen and turning it directly into protein, bypassing fertilizer, crops, animals—everything? Could that happen? It might happen with quantum computing, especially if we're going to colonize other worlds. There's a whack thought for you.

LINEAR VERSUS EXPONENTIAL GROWTH

I did not have my first Big Mac until I was sixteen. My first rock concert, pathetically, was the Bay City Rollers. A thirty-mile trip to the next big town to have Chinese food was a big deal when I was growing up. My high school graduating class boasted eighteen students. I grew up in a linear world where thirty paces were thirty paces.

Today we live in an exponential world where the first pace is one step, the second is two steps, the third pace is four steps, and by the time you have taken thirty exponential paces, you have circled earth twenty-six times. That, my friends, is the exponential world we live in today.

In the early stages, exponential growth is slow and sublinear even dipping into the negative, but at some point, it turns upwards and keeps on moving in a forward and upward trajectory—exponential growth looks like a hockey stick.

Look at smartphones. When smartphones were introduced in the early 2000s, very few people had them. By 2019, they're everywhere. That's exponential growth.

Newer agricultural technologies, such as genetic engineering, are also experiencing exponential growth. For example, when herbicide-tolerant corn and canola came out, few farmers tried it, but those who did had great crops. Other farmers saw their results and tried it the following year. The year after that, more farmers tried it and today, most farmers growing corn, soybeans, cotton, and canola are using genetic engineering technology. Bt brinjal (eggplant) is working so well at decreasing pesticide use and increasing yields that seeds are being smuggled from Bangladesh into India, where it is illegal to plant GMO seed. The adoption will be exponential, just like Bt cotton.

Exponential growth explains why technology hits and gets adopted so fast. We have the ability to build, and then share, technology that is beneficial. When something hits the farming community and works, the information spreads quickly.

DATA AND GREENHOUSE GASES

Today there is a lot of talk on the farm about data. How to capture it, how to track it, how to store it, how to leverage it, who owns it, and how to monetize it. And the amount of data coming off the farm is only going to grow.

From the dawn of mankind through circa 2003, globally it is estimated we produced five exabytes of data in written form (one exabyte is one million terabytes). In 2003, data production began to grow exponentially; by circa 2010s we were producing five exabytes of data every two weeks. Today we are producing five exabytes of data every two days.

The amount of data generated from the dawn of man to 2003 can now be generated in two days! And, we don't want things to change. Ha!

When you link the exponential growth of data with massive computing power, new ideas travel around the planet at rapid speed.

As I have shared, my interest in agriculture coincided with an interest in computers and technology. During my early Strategic Coach days, I dreamt about building a system that would deliver unbiased advice to farmers and, as previously mentioned, I founded the Agri-Trend Group of Companies. Our mission was to help farmers allocate scarce resources to produce a safe, reliable, and profitable food supply in an environmentally sustainable manner. Agri-Trend never sold fertilizer, chemical, or seed, but we developed the infrastructure to provide knowledge and wisdom to farmers through a coaching network. As Agri-Trend grew, we ended up with hundreds of coaches working with thousands of farmers on millions of acres.

To make all this work, we built an online platform, the Agri-Data Solution, that went live in 2001. It went into the cloud before there was a cloud, and we referred to it as "netware" because unlike software that came on 3.5" floppies or CDs, Agri-Data was 100 percent coded for online only.

Now that we had an online farming data system, what could we do with it? Was it possible to turn ones and zeros into dollars, we asked ourselves.

As I have discussed, as farmers reduced tillage, they increased their organic matter. And, for every 1 percent of organic matter, they would sequester or pull twenty metric tons of carbon dioxide out of the atmosphere. We thought if we could track and prove that, we might be able to sell it.

Our online data system allowed us to be among the first to monetize data and back in 2007, we founded Agri-Trend Aggregation, a carbon credit company. Agri-Trend Aggregation traded in the area of carbon offsets and worked with the government to write policy to monetize carbon stored in the soil due to reduced tillage practices by farmers. We were able to monetize that data and turn it into tradable fungible credits for industry and for farmers so that they would be able to trade.

In 2007 Alberta, Canada was the first North American jurisdiction I know of to implement a mandatory greenhouse gas system for its 104 largest final emitters. These emitters could either reduce emissions, pay into a tech fund, or buy offsets from someone else who was removing greenhouse gases.

The legislation created the motivation for farmers to reduce tillage, and we turned the data into credits to offset greenhouse gases from other sectors.

More on data later, but for now, let's talk about how data is converging with other technologies to reshape our vision of the farm.

CONVERGENCE

Today, we think of technology as computers, data, programming, robotics, or artificial intelligence. Technology, however, is a tool, and agriculture has always been about humans working with tools to grow the food they eat. A spade is a tool. A plow is another. A precise seeder is yet another. Each building on the next.

Each era of agriculture has been impacted by technological advancement. You cannot have one without the other.

In the muscle era, we figured out how to plant a seed. In the machine era, we figured out how to till the soil. In the chemistry era, we learned how to fertilize the soil while controlling pests and disease. And, in the biotechnology era, we improved the seed. Today, we're utilizing all these technologies together to again change the face of agriculture.

The technologies from the five eras of agriculture are coming together, allowing us to do things we've always done, but in a more efficient and precise manner because they are converging on the farm.

The muscle we're using today is our brain, instead of our brawn. Sensors are providing readings about what is going on in our fields. The machines we use are larger, self-driven, and self-regulating. The chemicals we use are safer and more specific. Biotechnology impacts are greater because of today's computing power.

Today, it is impossible to talk solely about one technology. They all mash together and they are mashing together at a faster rate.

This is convergence on the farm—this is the future of agriculture.

PART TWO

HOW WE FEED THE FUTURE

I have stated the next thirty years will put agriculture to the test. If we can feed the world in the next thirty years, the years that follow will, theoretically, be a cakewalk. The challenge to meet 2050 food demands while minimizing or reversing environmental harm from agriculture will be critical if we are to ensure that agriculture is infinitely sustainable.

There is no one right answer. Genetic engineering will not solve all the problems. Going organic will not solve the problems. Veganism will not solve the problems. Regenerative, agroecological, or any other buzzwords will not solve the problems. Agriculture is hugely complex, and there is no one solution.

However, I do believe technology will play a role in the future sustainability of our food supply regardless of farming method or geographical area. I get excited about this.

This fifth iteration of agriculture—technology convergence on the farm—will build on and combine the best from the previous four eras. The machines that appeared in Agriculture 2.0 are utilizing the GPS guidance and computing power of Agriculture 5.0.

We are on the cusp of driverless cars and on the farm, we have the mechanics and the electronics that allow the tractor to guide itself up and down the field. We have the ability to vary the varieties of seed we plant and adjust the amount of fertilizer we apply because computers can process the data and instruct equipment to make real-time changes.

It's a very exciting, albeit confusing time. This convergence is a time of rapid change. The rest of the book talks about where this convergence is headed, how fast things are changing on the farm, and how these technologies are going to impact farmers' productivity.

We've seen where agriculture has been, now let's look at where it's going—let's look at how we Feed the Future.

Chapter 6

BIOENGINEERING: GMO, FAKE MEAT, AND 3D PIZZAS

Perhaps the most important tool in the agricultural toolbox to help farmers feed the future in a world facing climate change and constraints on resources is bioengineering. This maligned science will impact all facets of our lives, from medicine to farming. It is important that we spend some time understanding the implications of bioengineering and embrace its possibilities.

ADDRESSING CLIMATE CHANGE

Most of us can agree the climate is changing. I won't argue about how much man's role is or is not contributing; I will leave that to the likes of environmental wizards like Dr. David Kawasaki. I can tell you that agriculture is affected by climate change. Globally, farmers will need to adapt.

In my career, I have watched as North American growing regions based on heat units have migrated steadily northward. Because of this, and augmented by genetic advancements, we now are growing corn and soybeans in Canada—unheard of at the beginning of my career.

Crops like corn ingest carbon dioxide and kick out oxygen. That's a good thing. And decomposed crop residue creates carbon sinks in the soil. The bigger the crop, the bigger the opportunity for increased sequestration. If we can engineer crops to do this better and faster, the environment benefits.

Biotechnology can help us deal with the effects of climate change. Whether crops are grown in hotter or colder conditions, they will need to be modified, so they can adapt to those changes. Mother Nature does this naturally, but this is a slow process, and our population is growing rapidly. Conventional breeding won't change fast enough to help us. We need biotechnology to breed crops that will adapt quickly to climate change.

For example, drought brought on by climate change results in high levels of sodium in the soil. The best way to tackle soil salinity is to grow crops in it, because when you grow crops in saline soils, they ingest some of the sodium from the soil. When you remove the crop, you take sodium off the land. Over time, you refurbish the land. But you must have crops that will grow in that saline environment to begin with, and biotechnology can help develop saline-tolerant crops.

Behind the energy from the sun, water is the most important input for growing crops and livestock, and agriculture uses about 70 percent of the fresh water we consume on the planet. It makes sense that we want to use water better. We agronomists refer to this efficiency as water use efficiency (WUE), and we work to measure WUE.

For example, we have measured that some farmers grow six to eight bushels per acre of canola on one inch of water, while the neighbor on the other side of the fence is getting ten to eleven bushels per inch of water. The difference is management that seeks to minimize loss and maximize the utility of the rainfall we receive.

Another approach to WUE is to manage the factors taking abiotic (weather) stresses out of the equation. Growing plants in a controlled environment is not new; greenhouses have been doing this for a long time, but creating a plant factory in an urban environment is (sort of) new. We see a great deal of attention being paid to new types of farming. Can these new methods help to feed the future?

THE URBAN FARMER

Given the fact that agriculture relies so much on water to grow the food that we consume, it makes sense that we look at new ways of growing food, especially high water-containing produce such as tomatoes or lettuce. I focus on broad-acre farming of staple cereal crops, but there's space for local urban farming, especially for fresh produce that has a high water content. We've already begun to see a shift toward urban farming or indoor farming in greenhouses. Urban farming can take advantage of the technological developments by growing high-water containing produce as close to the point of consumption as possible.

For example, every day, fifteen hundred trucks filled with produce leave Yuma, Arizona, and head toward the north and east of North America. Much of that produce could be grown closer to the consumer in urban farms. We don't want to ship water in the form of a tomato if we don't have to. Those farmers in Yuma reading this book are going to be upset with me, but this is a trend that's happening globally.

Panasonic has a huge indoor farming operation amid Singapore's urban sprawl. We're going to see more of that coming over time. The science we have today allows us to utilize LED lights to stimulate photosynthesis and grow crops in environments where those crops couldn't normally be grown.

It is important to also realize that not all sunlight is equal. The angle the light hits the earth is different in the Arctic than it is at the equator. Companies like G2V are working with spectral-controlled engineered light enabling us to grow vanilla in Edmonton, Alberta in the dead of winter.

ROB'S RANT

I do find it interesting how activists label commercial growers as "factory farmers" when, in fact, the indoor vertical farms are far more "factory" in nature, with issues such as light control, heating, and ventilation being major considerations. These in-door operations are, in fact, plant factories where conditions are tightly controlled while a three-thousand-acre broad-acre farmer must deal on the fly with the biotic and abiotic stresses Mother Nature throws at her. Which one sounds like a factory farm to you?

Combining environmental controls such as heat, light, and water is one way to increase efficiency. Another layer on top of this is using the power of genomics to increase agriculture output through synthetic biology, that would help us to respond to climatic challenges as well as produce more consistent, nutrient-dense food.

SYNTHETIC BIOLOGY

I was glad to get away from the cold Alberta winter for a few days in February 2016. California was in the fifth year of a drought—I know that made things hard for the farmers and golfers—but I enjoyed the sunshine of La Jolla for a few days. I'd be spending a few days with Dr. Craig Venter's team at the Human Nucleus Institute. Dr. Venter was one of the first geneticists to sequence the human genome, and now they were going to sequence mine.

I felt like I'd won the gene pool lottery. I was one of a cohort comprising a few hundred people to participate in the study that year. The Health Nucleus team would look at my metabolomics, genomics, and kinesiology, and then record what they found. They'd conduct 2D and 3D imaging of my heart pumping. And, most interesting to me, they'd sequence my genome. They'd share the data with me, but more importantly, our data would become part of a larger database that researchers would use to track genetic markers for diseases. If a new marker was discovered, they could compare that marker to my genome. If I had the same marker, they would let me know I was at risk for that disease.

You may be asking yourself, "What does any of this have to do with agriculture?"

Allow me to explain.

Genomic sequencing can now be done on any living organism: humans, animals, and plants.

One of the technologies that will most affect the future of food is genetic engineering, or what some are calling synthetic biology. The convergence of biological engineering with other farm technologies will enable us to grow more with less. Behind this technology is a new programming language.

Synthetic biology is the new language of programming in agriculture. Synthetic biology uses four letters: A, T, C, and G, which are the four key proteins that make up our chromosomes. The ability to program those A, T, C, and G lends itself to a whole new area of science that we'll have to grapple with, just as we did with the computer when it was new.

We all need protein. We must get this protein from plants or animals. Wheat, for example, is one of the staple crops providing protein to a great percentage of the human population. While we marvel at the ability to decode the human genome, wheat has been much harder.

Scientists have recently sequenced the wheat genome. This was a wicked accomplishment because the wheat genome has six billion DNA letters and is much more complex than human beings who have only three billion DNA letters (and you thought you were at the top of the food chain). This will now enable scientists to begin tackling issues associated with wheat, such as potentially adjusting gluten levels or biofortifying wheat with iron.

As scientists crack these genetic codes, they're able to search for markers of positive and negative characteristics just as the Human Nucleus Institute is doing for me using my genome.

A practical example of using biotechnology would be to create a wheat with the genetics to fight Ug99. You probably have never heard of Ug99, but this fungal rust, named after being discovered in Uganda in 1999, is one of the most devastating diseases in wheat. When it attacks, losses are 100 percent. Ug99 is moving around the planet and scientists are working to contain its outbreak. For example, Australia is on the watch for Ug99 and if it gets into the country, the predicted losses would be catastrophic.

With genomic sequencing, we're able to search for the characteristics that make a strain of wheat more resistant to wheat rust. Hopefully, we eventually will do this for Ug99.

ROB'S RANT

If we are going to have a shot at making agriculture infinitely sustainable while providing healthy nutrition for all people, bioengineering is likely THE most important tool in our kit. I implore you to reject the activist memes on social media and begin to dig into understanding this area of science better.

GENE REPAIR AND SUPPRESSION

Mother Nature has been editing genetics since life first appeared on the planet.

I had the good fortune to spend a week touring the Galapagos Islands with a group of friends. This area is a prime example of genetics constantly evolving. The one that struck me the most was a cactus on one of the islands. In Arizona, the prickly pear cactus has sharp needles. Don't touch them—they hurt. On one of the Galapagos Islands, however, the prickly pear had "hair" and no needles. Due to an absence of natural predators and because of exposure to adaptive radiation, this cactus changed to having soft, fuzzy hair instead of sharp needles.

Every organism on the planet is constantly mutating. DNA is constantly adapting, but those adaptations can take multiple generations and sometimes hundreds, if not thousands of years. Today, the earth's climate is changing faster than plants can adapt. Synthetic biology accelerates traditional breeding methods such as cross-pollination and hybridization.

With genomic sequencing and synthetic biology, we can harness the power of the computer to create crops that withstand the pressures of the changing climate. Whether climates turn hotter or colder, wetter or drier, or soils turn more saline, we can utilize technology that gives farmers the tools necessary to tackle those challenges in changing environments.

One of the most powerful biotechnologies converging on agriculture is called CRISPR. This is one to keep your eyes on; it is truly transformative.

Clustered Regularly Interspaced Short Palindromic Repeats (CRISPR) technology is like search, cut, and replace in a Word document. It allows for precise editing to DNA. It doesn't introduce anything new; it repairs a deformity inside your genetic code or clips off an area that's causing problems or simply not adding value.

While CRISPR is a great advancement on previous genetic engineered techniques, it still has some challenges. To alter DNA the CRISPR must be small enough to move through the pores of a cell wall. And it is, but it is also kind of floppy, and this might affect the accuracy of the system.

Enter nanotubes.

NANOTUBES

One of the latest advancements—just three months before I wrote this book—is the utilization of carbon nanotubes in genetic engineering. This is so cool!

Nanotubes are very tiny, perforated, rigid tubes on which one could embed genetic material like CRISPR for precise DNA delivery. Scientists put the desired trait on a carbon nanotube, which is small enough to slide through a cell wall. The cell then plucks the genetic material right off the carbon nanotube as if it were part of the cell.

Carbon nanotube technology, when combined with the other bioengineering tools, has the potential to dramatically increase the accuracy in which we can genetically modify synthetic biology. This could help us to engineer plants to be more input efficient, fight pests themselves, or be more resistant to cellular breakdown causing food loss.

Speaking of food loss, let's touch on that. Again, as a consumer concerned about the environment, I am open to looking at novel ways to reduce the losses experienced in our food system. Enter the topic of Arctic Apples.

ARCTIC APPLES

A few years ago, we had the opportunity to visit Neal and Louisa Carter on their farm near Summerland, British Columbia and got the story of Arctic Apples straight from the inventors themselves. During our three-hour visit, we cut up and munched on apples and they did not turn brown. WTH? This was so surprising. And better yet, nothing had to be added to the apple to achieve this amazing benefit.

You see, the apples had been engineered to reduce the browning. This was done through a process called gene silencing. Specifically, the polyphenol oxidase (PPO) enzyme, that causes apples to go brown, had been silenced.

When the apple flesh is exposed to air, the apple's genes activate the polyphenolic oxidase reductase enzyme, which is the enzyme that makes the apple turn brown. The researchers at Okanagan Specialty Fruits identified genes that cause the browning process and were able to turn them off, resulting in an apple that doesn't turn brown. They didn't introduce any new genetic material; they simply worked with what Mother Nature provided.

ROB'S RANT

Arctic Apple has nothing added to it. We have simply flicked off some genes that slow the browning process by about 90 percent (yes, the apple will eventually brown). Every year little Johnnies all over the world throw out their browned apples because they look ugly and taste icky. And, yet activist groups would have you believe that a GMO apple is bad. Is it? I hope you will now actively look for and buy, share, and eat the Arctic Apple. It is a concrete example of using science to reduce food waste.

Think about that. Consumers often discuss their concerns about food waste. Arctic Apple's technology potentially reduces hundreds of thousands of tons of wasted apples every year.

FOOD WASTE

When you are discussing food and agriculture, food waste always comes up. And we should be looking at ways to reduce food waste.

I am involved in an agricultural think tank associated with a group called the TWIN (The World Innovation Network) based in Chicago. This group has a wide range of experts around the table including the US Food Bank, PepsiCo, Griffith Foods, food scientists, and 2Blades Foundation and is organized by Dr. Robert Wolcott of Clareo—in other words, a diverse and eclectic group.

At our TWIN discussion, Diana Hortvath of 2Blades Foundation pointed out that while we in North America associate food waste with expired best-before dates or restaurant waste, in much of the world food waste happens in the field and in storage before it ever gets to the market or the kitchen.

On our farm in Uganda, losses both in the field from pests such as corn borer and in-storage losses from fungus and molds can be devastating. Again, we need people to understand it is the application of science that would reduce food waste.

A novel example of this is a company in Canada that is composting food waste from restaurants and grocery stores and then mixing the compost with elemental sulfur to make a fertilizer. Elemental sulfur must be converted to sulfate before it is available to the plant. This conversion is done by little bugs which eat the sulfur and poop out sulfuric acid (talk about red ring) which converts to sulfate for use by the plant. This bio-cycle product is spread on cropland at high rates. The bugs from the compost convert the sulfur over time to fertilizer for the crop. I have used this with my farmers providing multiyear nutrient availability with one application. Cool huh!

A closing thought on food waste. Maura Shea, of the Feeding America Group, cautioned on the issue of food waste by sharing another side of the story. She said her group tackles the challenge of feeding the one in eight Americans who are food insecure daily—meaning they don't have enough to eat. She said that as companies began to track and deal with food waste, some of the people and organizations that had been depending on this source (like ugly produce) saw their supplies dry up, which created even more food insecurity at ground level. There is always another side to the story.

NITROGEN-FIXING CORN

A flip side to agricultural waste is on the input side. The growing of crops requires inputs, otherwise known as fertilizer. Many of these inputs, like nitrogen, are not 100 percent utilized by the crop, meaning that some of the nitrogen makes its way into the environment. Can we use technology to tackle issues such as this?

Corn does not make its own nitrogen, but corn is a major user of nitrogen, and farmers supply nitrogen in the form of fertilizer. What if corn could make its own fertilizer? What if, like a legume plant, corn could be engineered to fix nitrogen from the air?

Would that be a good thing? Would you endorse this?

These types of examples are exploding in agricultural research. As mentioned in chapter 4, researchers have recently identified an indigenous corn in Mexico that can fix its own nitrogen. The farmers noticed a mucus-like substance oozing at the base of their corn plants, just above the soil surface. At first, they were concerned, but later noticed that the plants with the oozing mucus were healthier. It turned out that the bacteria inside the mucus was able to fix nitrogen. The researchers have gone off to study the genetic markers of that strain of corn.

And before you protest and say they'll rob the farmers of their special corn, know this: the farmers and the researchers have signed agreements stating that if something comes of the discovery, the farmers will receive a financial gain.

I think this is a great example of corn adapting to the environment by doing something corn is not supposed to do—nature is always adapting.

MOTHER NATURE IS A FIERCE COMPETITOR

Much as we praise Mother Nature, she never stops fighting, seemingly against us. She's constantly coming out with new diseases, insects, fungi, and molds. The pests that attack agriculture are constantly evolving to get around the strategies we create to control them.

You see, there is nothing natural about farming. We

constantly hear how monoculture is bad, but the truth is that farming is not in harmony with nature. Farmers in the Midwest US grow corn and soybeans—because they can—and generally, it makes them money. Farmers in Brazil grow sugar cane—because they can—and it makes them money. We grow canola in Canada—because we can—and it makes farmers money.

We cannot grow sugar cane in Canada, even though cane in our crop rotation might be nice.

More diversity in the crop rotation is a great idea. I work with many farmers who have five or six crops growing on their farm at the same time. Corn, wheat, barley, canola, flax, and peas or hemp might be crops we can grow one year; however, the logistical challenge of managing a wide variety of crops on a broad-acre farm can be daunting.

Some farmers are experimenting with intercropping where you plant two or more crops together and then separate the seeds after harvest; the jury is still out on this one. Crop diversity is a good thing and farmers are eager to move to multiple crops—if they can pay the bills and deal with the stress nature throws at them.

Mother Nature isn't always a blue-eyed blonde fairy with a flowing dress and flowers in her hair, floating through the fields spreading goodness. She provides much goodness, but when you're farming, you get into a lot of fights against her. She doesn't care about a bumper crop of soybeans. She's not interested in weevil-free cotton. She is interested in taking what she wants, by any means necessary.

We have to recognize that we're dealing with a formidable competitor, and we must have the tools to allow us to wrangle with the conditions Mother Nature sends, otherwise we won't be able to feed the people who we need to feed.

I've already received a lot of backlash on social media for my opinion of a fierce, competitive Mother Nature. Some of my naysayers cry that Mother Nature is upset because we're messing with her. As far as I can tell, we have always been in a dance with her. Adapting and manipulating at the same time.

One of the purposes of writing this book was to help provide you with context. To better understand that farmers are always battling the elements (unless you are farming in an enclosed box), and we must be able to employ scientific advancements in a way that maximizes energy in/energy out.

When it comes to energy in/energy out, meat often comes into the conversation.

TO MEAT OR NOT TO MEAT?

If a cow is alone in the forest and it farts, does it make any noise?

When it comes to meat, the biggest and loudest objection is, "Cow farts are killing the planet!"

This is not true. Cows do not so much fart, as they burp. They are ruminants, that is they have four stomachs, and—like their ruminant siblings, sheep, camels, and goats—cows are capable of bringing the partially digested grass back up for further chewing (chewing the cud). Methane (CH_4) is released into the atmosphere in a process known as enteric fermentation.

Every week I see some slide or meme saying livestock contributes 25 or 30 percent or more to greenhouse gases. The activists trumpet that livestock is killing the planet and causing catastrophic climate change. Again, I urge balance in the discussion; instead of getting memes from activist sites, read some science.

The initial data reported in the media was wrong and has since been corrected, but that, my friends, has not been picked up by the media.

The Food and Agriculture Organization (FAO) recently reported on the Intergovernmental Panel on Climate Change numbers regarding livestock. According to the IPCC numbers, global direct emissions from livestock are 5 percent, while transportation accounts for 14 percent. However, the IPCC then went on to report whole-cycle greenhouse gas emissions from livestock (including the raising of the feed, etc.) at 14.5 percent of anthropogenic (human activity), but the *IPCC did not do whole-cycle assessment for the transportation sector*, which would include the building of the cars, the acquisition and refining of the fuel, and the disposal of the automobiles.

So, it's comparing apples and oranges.

So, NO, livestock is not responsible for higher GHG emissions than transportation. In Canada, total livestock emissions are only 3.3 percent of the total annual greenhouse gas release, not 25 or 30 percent or whatever is on that PowerPoint slide designed to panic you.

ROB'S RANT

Every time I see "cow farts" on a presentation, I am going up to the speaker to ask them where their information comes from (likely Netflix) and I am going to correct them. I hope you do the same.

Finally, how come people are quick to point at cows, but have somehow forgotten that there were up to forty million buffalo burping in North America? BTW, us North Americans, we fart out about forty-three million pounds of methane annually. Maybe we should invent some human plug device to reduce our methane? Could be a big seller—and maybe we could each generate our own carbon credits?

Well then, if cow burps are not poisoning the planet, what then? Should we have livestock?

Well, globally, animal protein is very important to human nutrition. So, maybe we should leave the cows to eat the salad and we can eat the cows. You see, cattle are terrific at turning cellulose (plant material) into protein, and a vast majority of the land in any given country is not good for cultivation of crops. That land, however, is a good environment for grazing animals that turn the cellulose and hemicellulose of the grasses they eat into protein.

In western Canada, there are about sixty million acres of cropland, while there are approximately 140 million acres of marginal land that would not be suitable for farming, but would be suited for pasture. Worldwide this type of land is well suited for animal production, provided it is managed correctly.

What's more, if you have been reading that "regenerative" or "regenerative organic" (whatever that means) is the way to go, then you also accept that this type of agriculture endorses livestock and, in fact, needs livestock for nutrient cycling.

Livestock plays a key role in protein for human beings. I can hear you from here saying, "Well, you can just eat lentils." Of course you can, but you have to eat three and a half cups of lentils every day versus three ounces of beef, which means to get the required protein from lentils, you are going to eat a lot of calories.

It's easy for people who live in a climate like North America or Europe to propose that everyone become vegan. We must be careful about preaching from an idealistic pulpit. The environmental conditions in places like Mongolia are not great for crop growth. Removal of animal protein from everyone's diet is simplistic and naïve and ignores vast landscapes well suited for animal agriculture.

But, but, but, what about PETA and animals?

Again, my work is focused on broad-acre agriculture and cropping. However, in my entire career of working with cattle, dairy, hog, and chicken farmers, I have never met one who was intent on hurting their animals. In fact, just the opposite; well-cared-for livestock are happy, healthy, and profitable. The terrible videos we occasionally see on social media are outliers and not the norm. If you want to visit livestock operations and talk with real farmers about livestock care, let me know, I would be happy to set you up for a visit.

PETA is not going to drive me away from a T-bone; however, if it has you, the good news is you are going to have more choices going into the future. You might even be able to enjoy a BLT (sort of).

LESS BEEF AND MORE SALMON?

Eat less red meat and more fish, they say. Salmon is good for you; eat salmon.

We can all eat salmon, but what about the pressure on wild salmon stocks? Maybe instead, we can eat ocean-farmed salmon? No! Farmed salmon are bad because they are raised in the ocean and don't eat their natural river diet, so, you know, bad. Let's review: if we all eat wild salmon it puts pressure on the fish stocks, and we can't eat farmed salmon (I really still do not know why, but okay) so what about inland salmon?

Land-raised salmon doesn't touch the ocean, won't put

pressure on wild salmon stocks, and could grow at twice the rate as regular salmon, decreasing both the feed intake and greenhouse gases. What?!

Starting in 1989, a group has been working on a genetically engineered salmon that will grow fast, is safe, and tastes great. The Aqua Advantage Atlantic salmon contains a gene from its cousin, the Pacific Chinook, and a promoter sequence (DNA fragment) from ocean pout, enabling this salmon to grow year-round instead of seasonally like farmed or wild salmon. The result is a salmon that comes to market in half the time and has a feed conversion ratio of almost one to one!

If you love salmon and you care about the environment, this should be a winner!

VEGAN BACON?

Genomics gives us a new, comprehensive understanding of how things grow. For example, if you're a vegan and you won't eat meat because you're concerned about cruelty to animals, then would you eat bacon that was grown in a petri dish in a lab and didn't have an animal attached to it?

Meat, as well as organs such as the heart, lungs, and liver, can be generated through cellular replication. Maybe you don't want to eat lab-grown meat because you think that science is icky, but what if you need a liver transplant and a new liver could be grown for you without waiting for a donor? These are the exciting and challenging questions biotechnology requires us to think about.

Memphis Meats, SuperMeat, and Finless Foods are examples of companies employing a new science called "cultured meat," where meat is grown in a lab generated from stem cells. "No pig harmed in the making of this bacon!"

This science, called *tissue engineering*, is the same being used to help treat patients with muscular dystrophy. Stem cells are the seeds from which the "meat" is grown on edible scaffolds that give the product texture just like real meat. From just ten pig stem cells fed with plant substrates, you could grow twenty-five-thousand pounds of pork a month—now that's makin' bacon!

The same technology is also being explored to grow leather for the seats of your vehicle. Yep, that leather couch came from somewhere. We just don't connect the dots. This new leather-ish tech could be really amazing.

PLANT-BASED PROTEIN

Converting plants into plant-based protein is not new.

We have all sorts of plants that lend themselves to protein products: peas, lentils, chickpeas, and soy are used as raw materials for meat-replacement products. I recently ate a dish that tasted just like scrambled eggs; however, it came from mung beans. Interest in meat-replacement crops used to be limited to vegetarians and vegans, but in the last decade, they've entered the mainstream.

At another event I tasted not-shrimp. I mean it looked, tasted, and crunched like shrimp, but was made from plant-based material (mostly yellow peas).

Beyond Meat has garnered plenty of attention lately. When I look at this through the lens of a farmer, I might be glass half-full or half-empty. The ranchers are not enamored with it (disclosure: I have cattle), but the crop producers might have a different take on it. The main ingredients in the Beyond Meat recipe are water, pea protein isolate, expeller-pressed canola oil, and refined coconut oil, and the color comes from beet juice extract. If you grow peas or canola, this might be an interesting new use for your crop.

Another product garnering attention is the Impossible Burger, which has as its ingredients water, soy protein concentrate, coconut oil, and sunflower oil. The thing that gives the product a uniquely meaty taste is heme from soy plants—identical to the heme from animals, and this is grown from a genetically modified process.

ROB'S RANT

But wait, scream the activists, Impossible Burgers contain soybean and Beyond Meat contains canola and these are, are, are GMO and OMG this is bad, and we have to stop this! Really?

Maybe better to eat a cow?

Let's throw economics on top of food philosophy.

Beyond Meat costs nearly $12 per pound. Grass-fed beef can cost as little as $3.70. A pound of dry lentils, meanwhile, sells for less than $2 (the farmer gets paid about 40 cents per pound). If you have limited money, which products are you going to buy for your family?

I would say the important consideration has less to do with philosophy and more to do with nutrition. What are the attributes in the food you are buying and how do these meet your needs?

3D-PRINTED FOOD

I attended Abundance 360 in Los Angeles with Peter Diamandis. On the expo floor, I saw companies exhibiting 3D food printers. They were printing edible cookies and pizzas. One stand showed a video of a hospital in the Netherlands that prints food according to the needs of the patient.

The age of the Star Trek replicator has arrived. Think about a child coming home from school and saying, "I'm hungry. Mom, please make me a pizza." And Mom says, "Make your own pizza." The child then goes to the 3D printer in the kitchen, taps pepperoni pizza on the touchscreen display, and a pepperoni pizza is "printed."

What kind of ink is in the cartridges, you ask? The food ink inside comprises a series of substrates: oils, proteins, starches, and binding agents that combine to create the requested food. Okay, but where do the substrates come from? They are the components such as lipids, specific proteins, and starches (which come from broad-acre crops), so a part of the crop is utilized to generate the component that would be inside the ink cartridge of a 3D food printer.

Undoubtedly, 3D food printing will be in kitchens in the foreseeable future and it will change the way meals are produced. You might consider it processed food. Maybe it is. But it could also be a new healthy way to get the nutritive components you need. You may say you only want whole foods, but how can you be sure you're choosing the whole foods that are right for you?

How do you know the components in whole food best matches what you need or your individual body type?

The food ink will come from substrates that contain certain attributes. Let's have a look at that.

FOOD ATTRIBUTES

Every year, prior to planting a crop, you will find thousands of agronomists (I am one), hitting the fields on quads, in side-by-sides, or in pickup trucks to pull soil samples. We push a probe into the ground as deep as forty-eight inches, we pull out the probe with the soil core, and then split the cores into segments of zero to six inches, six to twelve inches, and twelve to twenty-four inches (you get the idea), bag the samples, and send them off to a laboratory where analysis is done on the soil.

I always request a complete analysis of the soil looking at the physical characteristics (organic matter, percent sand, silt, and clay). Chemical factors like pH and electrical conductivity, nutrients such as nitrogen, phosphate, potassium, sulfur, calcium, magnesium, boron, copper, manganese, zinc, iron, molybdenum, and occasionally cobalt and nickel are also analyzed.

With these numbers and with knowledge of the environmental conditions, we then formulate a fertility plan that will match the farmer's desired target yield based on the crop he/she is growing.

So, why share this with you?

Because, if you, as a consumer, want more copper in your wheat, we can do that. If you want higher levels of calcium or zinc in your food, we can do that. We can and do influence protein levels in wheat through nitrogen management. We can influence oil production by working with sulfur. Using agronomy, we can influence the attributes of the food coming off the field.

In most cases, there is no way for the farmer to get paid for crop attributes. This is because most of the production is on a commodity basis that primarily pays the farmer for the quantity of the crop produced, not the quality of the crop produced. However, there are a number of factors at work that are slowly changing this model.

Given what I just shared with you, would you pay more for grain grown under a system where complete soil analysis was matched by recommendations and application technology that resulted in the best, most sustainable application of fertilizer that ALSO resulted in higher nutrient density?

I think you might.

This is one of the areas currently being worked on. This ability to track how the food was produced; to segregate the grains produced under a specific regime, and then to tell you the story about how that crop was grown, comes down to data systems that can help to provide you with attribute-based foods.

Regardless of whether the crop was grown under conventional, organic, or regenerative systems or whether it was genetically engineered, a hybrid, or derived through mutagenesis, I think the important questions you should be asking are:

- Does this food have the nutrition I am looking for?
- Does the farming method support long-term sustainability?
- Do I get the information I need to trust what is being said on the label is true?

ANDI

What if we could grow the food in a certain way to provide you with a wheat that made a bread higher in selenium that might help to reduce prostate cancer in men? Would you buy that?

What would that look like? It might look like ANDI.

I am not a fan of Whole Foods (now part of Amazon)

marketing. I think they peddle fear to drive customers to spend "whole paychecks" based on their high-priced positioning; however, I do give them full credit for their implementation of something called the ANDI (Aggregate Nutrient Density Index Rating).

The ANDI food rating system developed by Dr. Joel Fuhrman rates foods based on the nutrition (like micronutrients) you receive based on the calories you ingest. The higher the score of zero to one thousand, the more vitamins, minerals, antioxidants, and phytochemicals per calorie. Leafy greens like the distasteful kale have a much higher reading than processed foods. Whole grains and dairy are somewhere in the middle.

ROB'S RANT

Consumers are often fooled by labels. They'll pay extra money for non-GMO spinach, when there's no such thing as GMO spinach—all spinach is non-GMO, so why pay more for the label?

Again, through the implementation of modern tools such as genetic engineering, I believe it is possible to grow crops, even broad-acre crops, that have higher nutrient density.

This may even be taken to higher levels or specificity (some might call it elitism) where foods are specifically matched to you, your blood type, your microbiome, your body—designer food for just you.

NUTRIGENOMICS

Nutrigenomics utilizes the same science we are using to engineer crops.

Based on an individual's genomic structure, the expression of genes could be altered by knowing how chemicals in food at the molecular level would react with a certain body type.

This could have profound implications for the control of obesity and for fighting diseases such as cancer.

As mentioned, I have had my genome sequenced, and now I am interested in following the science of nutrigenomics to find out if there are specific foods, nutrients, and/or chemicals that would be beneficial to my health.

This is a long way from going mainstream; however, with Moore's Law applied to genomic sequencing, it is possible for us to imagine a world where the majority of people have the information to better match the food they eat to their bodies, based on the attributes of that food.

BIOENGINEERING: GMO, FAKE MEAT, AND 3D PIZZAS

Chapter 7

PRECISION AGRICULTURE: SENSORS, ROBOTS, AND DATA—LOTS OF DATA

Because you have read this deep into the book, I can assume, you are concerned about how we are going to feed the future. How are we going to accomplish this while moving agriculture to infinite sustainability? Well, the answer is in your hand. It's technology; the same tech that goes into your smartphone is embedding itself in agriculture every day and it will do so at an accelerated pace into the future.

In these next sections, I'm going to take you from the present to the future of the ag tech world. I will do my best to help provide an overview of the technology that is shaping the farms of today and give you a glimpse into the farm of the future—an agriculture far more precise than the first four iterations we have discussed.

As near as I can tell, the first reference to "precision agriculture" is in a comic-type drawing of futuristic farms by Arthur Radebaugh. If you Google Arthur Radebaugh "factory farming" you will find an image that encapsulates much of what I will be discussing in this section. (Seriously, Google it—it's wild!)

The text accompanying the cartoon goes like this:

> Agriculture in the world of tomorrow will be so mechanized that farms will actually resemble factories. Crops and livestock will be raised on regular schedules under uniform and carefully controlled conditions. *Sensors,* those automatic control devices for today's wonder machines, will be adapted to the requirements of precision agriculture. They will take the place of human judgment in deciding and reacting to soil conditions, crop maturity, moisture levels, weather forecasting, feeding needs, etc. Bendix researcher W. E. Kock has reported that instruments to do this already exist or will soon be developed. The final part of the job for tomorrow's farmers will involve the packaging of the grown foodstuffs and their shipment to market—accomplished just as automatically as the growing itself.

This was published in 1961!

The science fiction of the 1960s is our reality some sixty years later. I could think of no better way to introduce this section of Agriculture 5.0 than to quote a visionary who could see the future that we now live in.

Precision agriculture has plenty of definitions, but to me, it means the precision application of crop inputs on microzones of the field to maximize crop output while ensuring environmental sustainability.

Those inputs could include seed, fertilizer, crop protection products, and irrigation water. Every one of these inputs costs the farmer money. The more we can correctly time the application of the right products to the needs of the crop, the better the efficiency. Remember the energy in/energy out equation? Well, precision agriculture simply takes this to a finer granularity. In fact, there are some that talk about varying the crop inputs down to the square foot! Can you imagine having the sophistication to vary fertilizer by nutrient down to the square foot!

How would we make that happen? What kind of machines would it take? What kind of information would we need?

Well, in order to apply something precisely, we must first understand what problem we are trying to solve. To do this we must measure and then turn those measurements into prescriptions or what we in the ag hood call *scripts*.

EARMUFFS FOR TOMATOES

"Are you using some kind of music therapy for the plants?" I said to the farmer giving me a tour of his state-of-the-art greenhouse. His tomatoes were wearing earmuffs like we'd wear during the Canadian winter to guard against frostbitten earlobes. He laughed a little and then explained that the muffs measure the mitosis of the cell division of the tomato and by doing so, they determine how quickly the tomato was growing so they could apply the proper amount of water and nutrients in real time to maximize the tomato's growth. In effect, the tomato was talking to the farmer.

Water is the number one nutrient that goes into the making of crops, regardless of the kind of agriculture we're talking about—indoor urban farming, greenhouse farming, or broad-acre farming.

Of course, plants need carbon dioxide and nutrients like nitrogen, but provided you have light, the biggest limiter of the photosynthetic process is water. The more efficiently farmers can use water, the better the conversion ratio into plant matter, and the better the profitability.

I'm most familiar with broad-acre, dryland farming, as

is done on more than seventy million acres in Canada and in places like Russia, Ukraine, or western Australia. The term dryland farming means that we don't irrigate and are exposed, and vulnerable, to what Mother Nature gives us in terms of rainfall.

We can't predict Mother Nature's generosity nor her scarcity, but we can measure and leverage the amount of moisture in the soil and in the atmosphere.

Years ago, farmers had simple rain gauges that measured precipitation by capturing the drops in a tube, and there was usually someone who had a bad knee or elbow who could predict the coming rain by the aches caused when the barometric pressure changed, due to an oncoming storm.

Knowing the amount of moisture in the soil is the first step in the crop planning process. We know it takes about four inches of soil moisture to build the crop factory. That is the main structure of the crop. After that, additional moisture goes in the development of the reproductive parts of the plant—the stuff that makes farmers money. We talked about WUE (Water Use Efficiency) earlier; this part of the discussion is about measuring it.

With a better understanding of how much moisture is in the soil, we can do a better job of predicting the potential crop yield. We could also predict how much more yield potential can be attained by the crop. This would provide the farmer with greater confidence in applying more fertilizer to attain some extra bushels.

ARE YOU SENSING SOMETHING?

Much like we measure our blood pressure or cholesterol to manage our health, farmers use sensors to determine the soil and plant health. Soil moisture probes can be buried in various regions of the field up to forty-eight inches deep and have a series of sensors at specific increments that can take electrical readings providing us with soil moisture tension. We can use this information to determine how much water is being held in the soil, and it will also tell us how far down in the soil profile the roots will need to penetrate to access soil-available moisture.

These devices are fantastic—but since they're buried in the ground, they're not much to look at.

In dryland farming, growers often underestimate how much water is left in the soil, and so they "give up" on the crop and stop applying inputs such as foliar fertilizer. If farmers have soil probes and are willing to trust the data, they can see yield increases by as much as 25 percent.

On irrigated farms, soil moisture measurement becomes even more important. Tying the soil moisture readings together with soil structure combined with information about crop stage and knowing the evapotranspiration rate—the combination of the water leaving the plant (transpiration) and water leaving the soil (evaporation)—we can vary the amount of irrigation water we are applying to the crop.

This process takes on an even higher level of sophistication if the farmer has a variable rate irrigation system that can adjust the amount of water applied to each part of the field based on soil texture. We can even use microwaves to help us with this decision.

Scientists have been using microwaves to determine soil moisture. More recently microwave sensors have been mounted on trucks and quads to gather information to develop soil moisture profile maps.

New sensor technologies are emerging that will provide information on soil nutrient status as well. Can you imagine not only knowing what the moisture status is but also getting a reading on the nutrient availability in real time, so you would know if the crop was boron deficient during a particularly dry spell? This would provide farmers with even more confidence knowing their application of foliar boron would boost the crop production and provide a return.

PREDICTIVE ANALYSIS

You think CSI investigative tools are sexy? Well farmers have been using DNA analysis to determine if disease spores are settling on plant leaves. Yes, farmers use DNA testing! Leaves from the field are picked and sent to a lab where DNA analysis will tell us if the leaves have been infested with spores of a disease that will harm the crop. This is still a pretty slow procedure, but it is gaining steam and will soon be more mainstream.

Another tool we use to help determine possible insect or disease pressure is on-farm weather stations. These in-field stations track rainfall, humidity, atmospheric and crop canopy temperature, solar radiation, wind speed, and other abiotic factors. Once the weather stations are tied together with disease and insect computer models, we tap into a whole other level of predictive analysis. What if, based on temperature and prevailing winds, the farmer knew spores of fusarium were likely to blow into her wheat? It would make spending that extra twenty dollars an acre on fungicide seem like a good investment rather than a risk.

Sensors are exploding in all areas of agriculture.

- Not too long ago I saw some plug-like sensors that were drilled through the corn leaf that was capable of reading the transpiration rate.
- There are sensors that can detect insect activity by

perfuming the air with pheromones (like Chanel No. 5 for bugs) that attract the insects to traps, so we can count activity and alert the farmer of increasing insect pressure.

- Recently I saw a presentation from a company commercializing a device that is taped onto the side of a mama sow. It is not uncommon for the sow to crush her own piglets by falling or sitting on them. The sensor being developed will pick up on the suffocation sounds and squealing of a squashed piglet alerting a caregiver who can rescue the little rascal.
- Potato storage sheds can be equipped with sensors that can help detect if the potatoes are beginning to rot so a farmer can turn the shed.
- Grain bin sensors are being used to monitor for carbon dioxide which could indicate grain bug activity. We routinely monitor grain bins for temperature to catch the grain before it "heats" and spoils.

Sensor technology provides information that drives resource allocation. The resource allocation depends on having the right information in the first place. And this is where technology is changing the face of the farm as we move toward convergence.

RESOURCE ALLOCATION—SCRIPT GENERATION

In our profession we are referred to as *agronomists,* but

maybe it should be *astrologists* because a lot of the time our job looks like fortune telling.

Precision agriculture is complex. It's like looking into a crystal ball (soil test analysis data), seeing through the smoke (yield predictions), and determining what the future will hold (making a recommendation or script).

Here's how the system works. We begin with the soil sample, which I covered in chapter 6, but now I'm going to take it up a notch or two.

Taking a soil sample should be an annual practice for farmers. Well before each planting season begins, a soil sample should be analyzed for nutrient and moisture content. The results of the soil analysis guide the farmer to choose the appropriate fertilizer, seed, and irrigation practices for the coming growing season.

Farmers and agronomists divide the field into grids or zones, so the samples are representative of the conditions of the entire field, not just a single area. They could divide the fields into zones based on topography, crop yield, soil or field conditions, or a simple grid.

For example, some agronomists work with grid sampling, which means taking a soil sample on every acre or two of the field and having all that data compiled into soil maps. Each sample is analyzed for multiple elements and a separate map is generated for each element—one for sulfur, one for phosphorus, and so on. After completion you might have twenty maps for twenty nutrients across the grids in just one field, at just one depth—that's a lot of maps.

Another method is to divide the fields into zones determined by previous yield maps of the harvest data, or according to a soil texture map generated by an electronic conductivity sensor pulled across the field, or by topography maps, or through the use of historic bio-vegetative images of the field over time that are layered together to give you biozones. The agronomist then goes to the field and soil tests in each of the representative zones.

Each part of the field has a target yield, and it's up to the farmer or the agronomist to program the equipment to apply different amounts of seeds and nutrients as the farmer passes over the field. For example, we may want to utilize two or three different varieties of corn, depending on whether the corn is grown on the top of the hill, on the side of the hill, or in the valley. We may have different corn varieties for north-facing versus south-facing slopes.

Based on historical yields and trend lines, we would determine what the potential target yield should be for each part of the field. This would result in a fertilizer recommendation (Rx). We would call for the right amount of nitrogen, phosphates, sulfur, calcium, and micronutrients to be applied in each area of the field. This is called *Variable Rate Application*, VRA, and the technology is referred to as *Variable Rate Technology*, VRT—taken together the VRA Rx goes into a computer on the tractor that controls the distribution of fertilizer and seed using VRT.

In the past, we took one soil sample per field and made one recommendation for various nutrients at one rate for the entire field. Now, we are looking at three to six zones in a field, each with different soil tests, and from that we must generate the scripts to vary each nutrient in each zone as we cross the field.

ROB'S RANT

The main muscle farmers use today is the brain. Some people have the impression that farming is a low-intellect occupation, and that just drives me insane! Farmers have to make more decisions on a daily basis during the growing season than any other business I know because of external pressures they don't control.

PRESCRIPTIONS FOR THE PLANTER

It is not uncommon to see up to five computer monitors in a cab of a tractor. One each for precision GPS guidance, seed operation, flow control, monitoring tractor diagnostics, and one for Netflix (seriously, I've seen that), plus a mount for your smartphone for tweeting.

All joking aside, the technology in a tractor, sprayer, or combine harvester would amaze most city folk.

As just mentioned, the VRA Rx (Variable Rate Application Recommendation Scripts) are transferred into the computers in the tractor cab in a special file format that will regulate the variable rate controller on the planter or the air seeder; it's called an "air seeder" because air from hydraulically driven fans is used to propel the fertilizer and seed through hoses all the way to the knife or disc which guides the product into the soil.

As the tractors move up and down the field pulling a planter or seeder and guided by GPS, an onboard controller distributes the amount of seed or fertilizer based on the specific GPS location of the unit and based on the VRA Rx for that part of the field.

Another key seeding and spraying development is something called *sectional shut off.* You can imagine as you come into a "V" part of the field, you would not want to sow in areas already sown. Again, using GPS, the unit knows where it has already seeded and can shut off the double application, lifting the shanks out of the ground as the seeder passes over that part of the field.

When a farmer spends $250 per acre for fertilizer and seed, eliminating a 10 percent overlap on a thousand acres saves her $25,000 (!) and reduces the environmental impact.

Precision agriculture is exploding because it helps farmers better allocate their resources. Instead of just throwing on blanket applications of fertilizer, farmers use precise amounts of fertilizers in exactly the right prescription for the right part of the field to get the yield they want. This flies in the face of the idea that the farmers are willy-nilly throwing a bunch of stuff on their crops.

As technology evolves, we will see on-the-go soil sensors capable of measuring moisture, nutrients, and soil attributes, such as compaction, in real time.

ROB'S RANT

Next time you hear someone saying that farmers don't know what they are doing with fertilizer and seed, share this story with them and then ask, who is the dumb one?

REMOTE SENSING THROUGH THE GROWING SEASON

As farms get bigger, it is more and more difficult to scout (walk the field) every acre. I know of one farmer in Illinois who is farming seventy thousand acres, meaning he would have hundreds of fields to monitor each day through the growing season. I work with many farmers that manage six, ten, or thirty thousand acres. We're not talking about twenty-acre hobby farms. We've got fields that are easily over three hundred acres each and I have been on single fields that cover more than twelve hundred acres!

In our region the growing season is short, about ninety to 110 days, which means the crop moves through growth stages quickly, so making decisions on the timing of product application is a real challenge. Farmers have to do this over large tracts of land in a compressed period of time.

We are increasingly employing sensors like cameras to help us manage the growing crop.

You are familiar with RGB (red green blue) cameras used on your phone. While these are useful sensors, we employ so much more in agriculture.

These are just a few of the remote sensors I have had some experience with—and more are being developed every year:

- Hyperspectral (a very wide range of narrow bands of light) sensors are expensive, yet thorough and provide big data, and can be used to ascertain plant stress, soil condition, growth rate, diseases—so many things, it's that good!
- Multispectral (select bands within the hyperspectral range) are a lower-cost option once you know what light bands are useful to focus on specific areas to attack specific problems.
- NDVI (normalized difference vegetative index) can pick up pigments in leaves to provide chlorophyll levels, which is a decent indicator of plant health.
- Thermal sensors pick up temperature; we know that plants under stress are hotter than normal, so these are pretty cool for picking up plant stress.
- NIR sensors capture light not visible to the human eye and can be useful to detect stress before you can see it.

This area of remote sensor technology is exploding, and each new advancement helps farmers identify issues in a crop in a timely manner; however we still need a way to "get" the picture.

DRONES

Robotics and drones come into play with this level of precision. As the crop is growing, we can fly drones over the fields to capture high-resolution, centimeter-by-centimeter images. The detailed images are then used to determine whether or not bugs or diseases are moving into a crop, and then use that information to target treatments.

As is so often the case, the technology is ahead of regulation. In North America, the Federal Aviation Authority (FAA) in the United States and the Canadian Transport Regulatory Commission (CRT) in Canada have regulations that make it difficult to fly drones.

At the present, a drone must be "tethered" to a human being that can monitor the drone with line of sight, which reduces the ability to gain efficiencies with drone technology.

As these regulatory issues are solved, we will see drone systems "catch and release"; that is to say they will wake up, fly their missions independently and land back ready to upload the data they collected in the field.

In Japan, drones are used to do the analysis and are routinely used to spray crop protection products on fields. This is especially suitable for smaller fields and/or for terraced terrain.

ANOMALY DETECTION

Every day through the growing season, satellites are

taking pictures of farmers' fields. Whereas drones might have centimeter-by-centimeter resolution, the satellite resolution we can obtain would be at best, one-meter by one-meter resolution, which is still pretty good. The cost of satellite imagery can be more reasonable than the cost of a drone; however, the Achilles heel is cloud cover, which can delay the receipt of a clear in-season picture.

Another strategy is to use aerial imagery taken by planes, which reduces the cloud issue.

Intelinair is an example of a company employing aerial imagery of the field taken several times during the growing season for comparative analysis. Planes fly over the field throughout the season, the images are stitched together, then images taken on different dates are layered to create a complete picture of the field over time.

Anomaly detection is then used to compare those images throughout the growing season to determine which part of the field is falling behind and which part of the field is surging ahead. When a problem area is pinpointed, the farmer can then scout the field to determine why that part of the field is falling behind or moving ahead.

It's more important for a farmer to know where a crop is changing than where the crop is good, and anomaly detection does that. As the farms get bigger and bigger, anomaly detection—either with drones, satellites, or aerial imagery—is critical because the farmer can't be everywhere at once.

Besides problem detection, we can use in-season images to build real-time sprayer prescriptions.

Data collected from in-season bio-vegetative maps can be used to precisely distribute pesticides through sprayers based on potential disease pressure and/or crop growth; we can VRA Rx pesticides to put what we want on at the right rate to solve the problem we see at a specific location.

Sclerotinia stem rot is a disease that attacks canola. When it attacks, it can drop yields 20 to 80 percent. Using in-season maps we can program the variable rate sprayer to apply the fungicide only on the parts of the field with the heaviest canola growth, and zero fungicide where the crop is thin or nonexistent. To watch this in action is jaw-dropping. The sprayer moves along at ten to fifteen miles per hour and its nozzles switch on and off depending on the crop growth. It is mind-boggling to see the nozzles shut off as the boom passes over a standing body of water. That's technology making an environmental impact!

SPRAYERS THAT SEE

We've established that it costs time and money to spray herbicides to control weeds in the crop.

It's possible to put sensors on the sprayers that can actually see the weeds. This is an area being pursued by many companies.

Photometric sensing technology would be capable of seeing a weed as the machine crosses the field and targeting the spray directly on that plant without having to blanket spray the entire crop. Can you imagine the savings if we could just take out the weed or target insects in specific parts of the field?

Self-driving cars employ a technology called LiDAR (Light Detection and Ranging) where the device fires a light pulse at surfaces up to 150,000 times per second and measures the speed of bounce back to determine shapes and sizes of obstacles it's approaching. We are using this technology to help robots see in the field.

You might not think of Bosch in agriculture, but they are developing some incredible technology whereby sensors can spot, identify, and spray a single weed in the growing crop while the sprayer is passing over the field!

GPS/GIS

Global Positioning System (GPS) guides farm implements in the field. Depending on the type of GPS system used and the type of accuracy correction devices, a 120-foot sprayer can traverse the field with sub-inch pass to pass accuracy.

When GPS coordinates are linked with Global Information Systems (GIS), every object and activity can be tied to a specific location in the field.

Sometimes before the crop emerges, farmers drive up and down the field hunting for rocks that, if ingested, would damage a $750,000 combine. Rock picking is a dull, dirty job.

Today when a person working the field sees a rock, they drop a GPS pin in their smartphone app to mark the location of the stone. This dramatically increases the efficiency because later they simply drive their rock picker from one known area to the next.

All of precision agriculture is tied to GPS and GIS. And this technology is part of the convergence going on inside of self-driving cars and, of course, field robots!

ROBOTICS

While 97 percent of farms are family-owned businesses,

many of these farm families struggle to bring their children back to the farm. The children that do return are true treasures.

Without a doubt, the number one challenge on most farms today is labor. Farmers simply cannot find enough qualified operators to run the advanced machinery on most commercial operations today.

Remember, farmers live in remote areas and face the challenge of finding qualified operators to drive a six to seven hundred thousand dollar piece of equipment. The challenge to keep them living in a remote area is greater still.

Often the only operators a farmer has are those retired farmers whose land was purchased by the current operation. This is good, but they are aging—many operators are in their 70s and 80s—and many struggle to keep up with the technology in the equipment.

Robotic farming is inevitable and there is hope that building a robot fleet for the farm will bring back some of those bright and talented minds to the farm that would otherwise be gone.

On the farm, robots can help farmers with three main issues: labor, safety, and cost.

Few young people are enamored with the idea of sitting weeks on end in a glass cage of a tractor monitoring computer screen activity and refilling bins with seed or fertilizer.

Today, thanks to GPS, tractors, sprayers, swathers, and combines drive themselves. Tomorrow we may see farms where allocated fleets of robots cross the land to do what was once a really dull, dirty, and potentially dangerous job.

For example, we're starting to crack the code, so we can utilize robots to pick up the grain from the combine and take it to the truck that goes into the grain bin.

As the grain is augured into the grain bin, a sampling device like the Ingrain system from Intelliconn can automatically divide and seal the samples in secure, identity-preserved bags that not only provide an accurate source of the grain, but also can be used for definitive attribute analysis.

Automation like this will ultimately provide more accurate data that can show consumers where the grain came from and how it was grown.

Several companies are pursuing a robotic strategy I refer to as the "headless horseman," which is a trac-

tor without a cab hooking that autonomous tractor to
existing farm machinery.

How long will it be before it is possible to buy an autonomous kit for an existing tractor, retrofitting your old
horsepower, hydraulics, and rubber and turning it into a
self-driving tractor? I was at a Trimble conference in Las
Vegas where I sat in a Case tractor that had been outfit-
ted to be 100 percent autonomous. I was allowed to ride
in the cab as long as I didn't touch a thing. The experience was surreal; I sat in the tractor while it crossed an
obstacle course, turning all the corners, slowing down
for obstacles, speeding up on straightaways, and repeating every circuit in exactly the same way. A fake truck
was parked on the side of the "field" to illustrate how
the tractor sensed the obstacle, slowed down to pass the
truck, then sped up again. All I had to do was sit there—a
long way from my dusty days on the Field Marshall!

Many companies are pursuing robotics in greenhouse
operations that can pluck tomatoes. There are photometric sensors that determine the perfect ripeness for
picking a fruit off the tree or vine. Robots change the
way that farmers configure their greenhouses or their
exterior growing spaces. They determine the most efficient way for a robot to get up and down the rows before
planting.

There are lots of companies chasing the small robot ag

space. This book is about broad-acre farming. So, what does robotics look like for the farmers of consequence? Meet Dot.

THE STORY OF DOT: AUTONOMOUS BROAD-ACRE ROBOTICS

A farmer in Saskatchewan wakes before the sun comes up, as he's done for the past forty years. He stumbles downstairs to the kitchen, pours himself a cup of coffee, and goes to the computer screen in the adjacent office to check on the robots working in the field. The robots have been planting the crop all night long. And the farmer is anxious to see what the progress has been while he was sleeping.

This is not a scene from a sci-fi film. There are no aliens creating crop circles in the wheat field. Autonomous robotic farming is closer than you think.

In all transparency, I'm not 100 percent unbiased. Since the beginning of 2019, I've been playing a role in the development of autonomous robotic farming for broad-acre agriculture; I'm the CEO of Dot Farm Solutions, the retail arm of a robotic-powered farming platform called Dot. (www.seedotrun.com)

Saskatchewan farmer and inventor Norbert Beaujot founded the company SeedMaster, which builds some of the largest seeding equipment available for farmers. As the fields expanded and the machines got bigger and more sophisticated, Norbert thought—as do most inventors—that there's got to be a better way to approach farming and farm machinery.

In the case of Dot (named after Norbert's mother, Dorothy), he designed a U-shaped power platform that connects to a variety of farm implements. Dot is a completely robotic autonomous platform, capable of running in the field and performing, as mentioned, a variety of operations.

It's called a platform because many devices can connect to Dot. Dot is like the body of a Swiss Army knife and the Dot Ready Implements that attach to Dot are like the accessories inside the knife. The farmer will choose the types of accessories he wants depending on his needs.

Dot isn't the only company developing robotic farming, so we are not alone in this category, although of course I believe our robot is the smartest, the prettiest, and for sure, Dot is the biggest. Dot offers a glimpse of how broad-acre farming will look in the future.

At the present, Dot is the largest robotic agriculture equipment in the world. She has four hearts—four hydraulic motors—and stands on four legs in the form of four independent hydrostatically driven wheels. Her Cummins diesel engine provides up to 175 horsepower to glide across the fields.

Dot represents the convergence of all the technologies enabling farmers to allocate resources, such as seed, fertilizer, and crop protection products, while reducing labor costs.

Before you get your panties in a bunch about Dot taking away jobs, let me tell you: farms are hurting. As I mentioned earlier, most farms are family businesses, but many of the younger generation aren't interested in sitting in a tractor cab all day, driving monotonously back and forth across acres and acres of field. They are interested in technology; figuring out how to get an autonomous fleet working on a farming operation is both challenging and fun.

Let's imagine what the farmer looking at his computer screen did the day before to send his robots out in the field.

First, he programmed the field layout for the robot, so Dot would know where to go. He consulted a GIS (Geographic Information System) map of the field to draw a geofence, or outer perimeter, around the area that Dot would travel. He'd highlight any obstacles in the field such as trees, lakes, rock piles, or fence lines, so Dot would go around them. He'd also identify any gullies or rough terrain, known as "school zones," so he could program Dot to slow down.

Unless the field changes, the farmer won't need to reprogram the field layout. If an unidentified object appears in her path, however, Dot's eyes (cameras, radar, and LiDAR sensors) would constantly scan the path and let Dot know if there's something unexpected in her area. And if there was, she would stop.

Say the farmer was seeding a new crop. He'd consult the agronomic information gained from the soil sample. He could program Dot to plant the quantity and variety of seed that's most appropriate for the topography and microclimates in the field. He could, of course, also add appropriate amounts of fertilizer for that type of seed in that soil and microclimate.

He would dock Dot with the seed drill or planter, load it with seed and fertilizer, and send Dot on her way to work for the evening. While the farmer attends other logistics of the business or his kid's basketball game, Dot does the work in the field.

All of these things would allow a farmer to be far more efficient at planting a crop. His safety increases because he's not out in the field driving heavy equipment, while his cost of equipment per acre decreases.

The farmer, from the warmth and comfort of his office, would be able to see what Dot sees through Dot's photometric sensors that are connected to the farmer's computer screen.

When Dot's finished seeding, she might disconnect from the air seeder and get ready to put crop protection products on another field by connecting to a Dot Ready Connect Sprayer. The agronomy team working with her would have provided the prescriptions and maps, so she sprays in precisely the right spot.

Robotics allow for a precision that human-driven tractors find hard to match. Autonomous sprayers equipped with all the latest technology will enable farmers to protect their crops at even lower costs with less chemistry applied to the field. These sprayers can do sectional shutoff or turn compensation (more spray on the outside of the turn, less spray coming out from nozzles on the inside of the turn) and can vary the amount of spray by individual nozzle to maximize productivity. GPS technology guides all the operations. Talk about accuracy!

Or Dot can go organic. She might be connected to a cultivation unit that's capable of helping organic farmers control weeds through the growing season by precisely culling the weeds in between the rows of the crop.

As time progresses, rather than feeding information

into Dot, she would be equipped with on-the-go sensors that read the soil nutrient levels in real time while Dot is moving through the field, and then, utilizing an algorithm (we'll talk more about this in a bit), change the fertilizer and seed in real time to match the nutrient levels. Chlorophyll sensors will provide the information Dot needs to adjust nitrogen application in real time.

Dot, and other robots like her, are a convergence of precision agriculture with data collection and algorithms. Dot ingests data from a wide variety of sources and pulls the data together through the course of the growing season all the way through harvest, as Dot's role switches from growing the crop to harvesting it.

Dot would assist in picking up the grain from the combine and bringing it back to the edge of the field where it could be put into a truck. She could potentially play a role in the yard by moving the augers that lift the grain into the grain bins on the farm. At the end of the auger, a sensory device could conduct spectroscopy analysis of the grain in real time, dividing the grain into even samples and informing the farmer of the grain's protein content or any other measurable attribute.

The data collection throughout the crop's life cycle from soil analysis, to seeding, through harvest, creates a data continuum by pulling together disparate pieces of data throughout the growing cycle.

AUGMENTED REALITY

Here the fun really begins. The data drives the machine, but the farmer can use the data, too. Think of putting on a pair of glasses and walking out of your farmhouse. As you look at the sky, information about the weather appears on the augmented reality device. When you walk into the field, you could look at the soil and, because the soil information is tied to a data system tied to the augmented reality glasses, you could see the soil information in your glasses—without getting your hands dirty.

You could then look at the crop and see the variety, the seeding date, and target yield. You could look at the leaf and, provided you had the information in the data system, it could give you the nutrient status of the leaf, it could tell you how much nitrogen phosphate, zinc, boron, and copper was in the plant and whether any of these nutrients were deficient.

You could look at an insect in the crop, identify it, and see the prescription for how to control it.

Augmented reality would allow you to walk over to your grain bins and see inside to view the inventory you have and the grain's attributes. You'd be able to look at your pumpkin crop and see the growth rate to make sure your pumpkins will be plump for Halloween.

Augmented reality is about connecting our devices

to data systems to gather information about what's happening on the farm in real time. Farming is a hard, often dirty, business. Interacting with a smartphone with dirty hands or with gloves is not an easy scenario. Augmented reality devices that are hooked up to data systems would be a way to enhance how a farmer interacts with data while the crop is growing.

VIRTUAL REALITY

A piece of farm equipment can cost as much as a small plane, even more than the average house; so naturally farmers like to see what they're buying before they sign the checks. Because of the size of the equipment, however, it's challenging to take a sample machine to the farm; the alternative is to fly the farmer to corporate headquarters, which is expensive, too.

Virtual reality provides a way for farmers to experience the equipment without leaving their farm. By putting on a pair of VR glasses, they could walk in and around a large machine, such as Dot, and see it working a field. They'd be able to understand its size in relationship to their own. They could switch between the electronics, the structure, and the hydraulics of the Dot platform and see each one of these systems.

The virtual visit accomplishes a few things. One, time and space shrink so that the farmer gets more comfortable with the unit quicker. He has a clear understanding of perspective and how the unit looks and feels, and potentially operates. Two, it provides a way to train people. And, three, provided it was hooked up with the sensors on the machine, virtual reality could help diagnose, service, and repair the machines as needed.

I bet you never thought we would be using the words "farmers" and "virtual reality" in the same sentence.

THE INTERNET OF FARM—CONNECTIVITY

Can you imagine running a five or ten-million-dollar business without the internet? Yet, this is reality for many farming operations off the main broadband grid. To have a smart farm we can't have a stupid internet connection; it just does not work.

Thankfully, as we stare into the future, we can see ubiquitous connectivity on the horizon. Someone, either Bezos, Musk, or Zuckerberg, is going to pull together the resources to make global 5G connectivity a reality.

This will accelerate what we call the Internet of Farm where many devices and sensors are constantly connected to centralized databases.

In the near future you could imagine the entire farm connected to a Wi-Fi mesh or a Wi-Fi dome so there would be no need for cellular SIM cards. We have this at our place. We have an outdoor Wi-Fi mesh that connects the house, the shop, the garden, and the pasture all with Wi-Fi so any camera or device we want to hook up uses a simple IP connection.

The soil probes left in the ground need to have connectivity to gather soil attribute (nutrient and moisture readings) data and send it back to a computer that compiles and crunches the data to provide the farmer with an actionable result.

Eventually we will have ubiquitous 5G connectivity that will enable sensors to collect data, make decisions, and deploy robots to make the applications.

What if we didn't even have to have sensors? What if the plants were the sensors themselves? You could have an Internet of Plants. In fact, this is being done where scientists have connected lemons to the internet using an FM signal. The plant determines whether it is wet or dry and sends a signal telling the farmer their situation without any soil probes at all.

The Internet of Farm is about bringing all these pieces together. It is happening on farms everywhere at increasing speed.

WHAT WILL THE FUTURE BRING?

If I was asked "what would make the biggest difference to the poorest farmers?" it would be a soil test. If we could digitize that process, we could democratize the information and eventually demonetize the systems to a point where every farmer on the planet could have the information they need to farm more profitability and sustainability.

Convergence will bring about advances to make this goal a reality. For example, what if the nutrient results appeared directly on the soil probe rather than having to send the sample to the lab? And what if in addition to nutrient and moisture content, we could read the pH, the acidity or alkalinity of the soil? Or, better yet, what if I could leave the soil probes in the ground and receive the instantaneous readings about nutrients, moisture, and pH directly on my computer in my farm office? I could then use the information prior to planting for the season as well as during the season to observe fluctuations and adjust my irrigation or fertilizing plan.

That'd be pretty terrific.

Let's take this one step further. Let's put a soil sensor device on my seeding equipment so the soil is analyzed in real time as the seeder drives itself across the field, analyzes the soil, receives instantaneous results about nutrient and moisture levels, and the onboard computer makes a real-time decision about how much fertilizer to put in the hole with the seed.

That would be really futuristic, but that would take *ALGORITHMS.*

Chapter 8

DATA COLLECTION: MACHINE LEARNING, ARTIFICIAL INTELLIGENCE, AND ALGORITHMS

On May 24, 2019, IBM announced global expansion of its Watson decision-making platform for agriculture. AgroNews[3] reported:

> IBM announced the global expansion of Watson Decision Platform for Agriculture with AI technology tailored for new crops and specific regions to help feed a growing population. For the first time, IBM is providing a global agriculture solution that combines predictive technology with data from The Weather Company, an IBM

---

3   "IBM announces global expansion of its Watson Decision Platform for Agriculture," AgroNews (May 24, 2019) http://news.agropages.com/News/NewsDetail---30573-e.htm.

Business, and IoT data to help give farmers around the world greater insights about planning, plowing, planting, spraying, and harvesting.

By 2050, the world will need to feed two billion more people without an increase of arable land. IBM is combining power weather data—including historical, current, and forecast data, and weather prediction models from The Weather Company—with crop models to help improve yield forecast accuracy, generate value, and increase both farm production and profitability.

The average farm generates an estimated five hundred thousand data points per day, which will grow to four million data points by 2036. Applying AI and analysis to aggregated field, machine, and environmental data can help improve shared insights between growers and enterprises across the agriculture ecosystem. With a better view of the fields, growers can see what's working on certain farms and share best practices with other farmers.

The future is now, and it's on the farm.

DATA COLLECTION

Farmers depend on two types of data collection:

- Measurements, which provide quantifiable informa-

tion about weather, soil, field, and crop conditions
from sensors or drones

· Recognition, wherein the sensor or drone captures
the image and the computer then recognizes the
condition and recommends a solution

We can't make sense of the information and convert
it to a real-time recommendation without a powerful
computer programmed with algorithms that understand the data. For example, coupling the nutrient
status with the target yield and then calculating how
much zinc to put on the field while crossing it. I find
these scenarios fascinating, but I also know the challenges farmers face.

The amount of information being ingested by a combine while harvesting a crop is potentially terabytes
per day. The combine records the amount of crop, the
yield of that part of the field, the grain moisture and as
sensors drop in price, we could track the attributes of
the grain as it's being harvested. One of the latest addi-
tions to combines has been on-the-go protein maps. As
we're harvesting the crop, we can get a visualization of
changing protein complex inside the crop while we're
moving across that field. This protein analysis will not
only provide you the consumer with information, but
it will also provide me the agronomist with an understanding of how hard the crop has pulled on the soil
nitrogen reserves.

The data is only helpful if you analyze it and use it to make decisions.

As sensors grow, so does the amount of data. Data is only good if you do something with it. More data, without systems to manage it, might create more problems than it solves.

DATA MANAGEMENT

We all face data management problems—remember those daily exabyte numbers—and farmers are no different. In this section, I will be talking more to farmers; you are welcome to eavesdrop.

The biggest issue we have with farm data today is the human entering the data. Every time we count on someone with big sausage farmer fingers to enter data into a smartphone or punch figures into a computer using a keyboard, we have a problem. The data systems are only as good as the humans entering the data, so we have human flaws and human errors.

As the data collection shifts to direct input from sensors, or live feeds from equipment, the data will get cleaner and bigger, and then we can use it to make better decisions.

For example, the yield maps that are generated from

combines while harvesting the crop aren't developed by humans. The yield maps are developed on the fly while the farmer drives the combine; the computer on the combine collects the data and develops the yield maps. Ultimately, it could be a moisture map or a protein map, whatever the sensors have been programmed to measure.

We are talking a lot about data, but we have not talked a lot about the data platform. I believe one of the most important decisions a farmer will make is what data platform will he/she work with. As mentioned, our Agri-Data Solution, now Trimble Ag Software, is an example of an agriculture data platform. This is one of a dozen or more farm management systems available to growers.

ROB'S RANT

A caution to the farmers reading this section. There is no silver bullet. No one data platform will likely do everything you are looking for, and data without applied science is garbage.

Poor agronomy plus precision agriculture equals poor agronomy precisely applied! Remember that.

Until the human hand is eclipsed by the machine making decisions based on data from the sensors, the process will continue to be highly manual and imprecise.

Over time many of these data companies will migrate their thinking toward the development of algorithms that can tell the equipment what to do. Those algorithms will be based upon past experience. Algorithms will come in a wide range of choices.

ALGORITHMS ON THE FARM

Farmers will have a wide variety of algorithms to choose from because two agronomists can look at the same field and recommend different scripts. You have to consider the risk tolerance the farmer is willing to take on, the likelihood of success from applying a specific type and quantity of fertilizer to the crop, and an understanding of the environment and the weather. The markets play a role as well.

Here's an example of a human agronomic algorithm at work.

I am going to put a soil analysis in front of a couple of agronomists and say,

"Program this field for a hundred bushels of hard red spring wheat, with 13.5 percent protein that will be grown near Saskatoon, Saskatchewan. The rainfall through the growing season is estimated to be ten inches, and we've got five inches of soil moisture right now through the course of the winter months."

Each agronomist would have a slightly different recommendation based on their biases. Agronomy isn't a black and white science; it's science tied to art and it has a bias based on the artist (or agronomist) looking at the data.

Likewise, the algorithm from one data platform will be different than the algorithm from another data platform because it's based on the decisions entered from different human experiences. Who's right?

Eventually, we will know whose script brought results closest to the yield target and the protein target. But we won't know that until the end of the season. In most cases, broad-acre agriculture has only one growing cycle per year. A farmer who begins farming at the age of twenty-five has only had forty crop cycles by the time he's sixty-five to master his trade.

Can you imagine if you only got forty shots at doing something on which your entire livelihood depended?

Farmers don't get to test and iterate like the tech entrepreneurs releasing new products every eight months or eight days. Farmers have to wait months for a crop to bear fruit, sometimes years depending on the crop. They don't know the outcome of their decisions for months after they make those decisions. And with the myriad variables out of their control, even repeating the same practice one year to the next doesn't guarantee the same results.

Since the development of scripts and algorithms come out of the data sets that the farmer uses, the choice of which data sets to use is a critical decision. I believe the farmer should own his data, but building this into the algorithm of a data system to feed information into a robot is a complex issue.

DOES YOUR DATA TALK TO YOUR EQUIPMENT?

And, the data systems need to talk to each other. Does your green equipment data system play nice with blue, red, and yellow equipment and vice versa?

A farmer can choose to align with an input manufacturer, meaning they choose a data platform that's provided by the crop input manufacturer, which could be a seed, crop protection, or fertilizer company. Do you want those companies to have access to your data? Maybe you do; maybe not.

Alternatively, you could go to an independent data company to manage the data on your farm. The positive side is that they're not tied to the sale of crop inputs or equipment; the downside is that independent data platforms are more expensive.

If the platform that you choose to manage your farming operation is a "free" platform, then undoubtedly you, or your data, are the product. You're being utilized by that company to do what they need to do, which can be fine—you just need to be aware of it.

After you decide which type of data platform to use, you have to start ascertaining the attributes of each one in that category. Some data platforms are stronger agronomically, while some are stronger on the marketing side, on inventory management, or enterprise management. Some talk to the equipment directly, others don't. If you want to work with two data platforms, they may be incompatible with each other. You want to find a clean data flow between platforms and equipment.

Farmers are facing these challenging complexities today. Slowly, like all new developments, the data systems will start to harmonize where data can flow regardless of the color of the equipment or data platform. Remember, in the beginning, Apple and Microsoft didn't talk to each other, and now you can share documents with ease. (Google Calendar and Microsoft Outlook Calendars still don't play nice.)

ARTIFICIAL INTELLIGENCE

Eventually the data gets so big you will need algorithms, and those algorithms will be adjusted based on artificial intelligence.

The algorithms are built through the process of ground truthing.

Ground truthing is a term we use for training a system to recognize something by showing it the truth from ground level. In the case of the augmented reality glasses recognizing a weed, its ability is predicated on the ground truthing of that weed. It may take ten thousand images before the machine learns to recognize a weed. Keep in mind that as the plants grow, they go through different growing stages and those growing stages change the look of the plant. The machines face the same challenge with insects, too.

For artificial intelligence to work a tremendous amount of algorithm development needs to be done, which is currently heavily dependent on regression analysis. In other words, we have to enter many, many images of the insects or the weeds that we want the machine to ultimately diagnose. The more images the machine memorizes, the more likely it will be accurate. Initially it's heavy human programming that is causing the machine to understand what the weed looks like or what the insect looks like, but after a while, the machine actually begins to learn on its own.

For farming and agriculture, the opportunities are great. No farmer wants to spend more money on pesticides or fertilizer than is absolutely necessary. Farmers want infinitely sustainable agriculture. The ability to recognize the weeds, the insects, the diseases, and other challenges that are facing the crops in the growing season would allow the machine eventually to make a decision on the application or the treatment of that particular pest that you're after in the field.

USING LIGHT TO TEST BARLEY

We hear about something called machine learning. What is that? And how do machines learn?

Let's look at barley. When you're making beer, you want malt barley with moderate protein levels. When you're feeding cattle, you want barley with a higher protein level. Malting companies that buy barley for beer are always testing for protein levels.

We wanted to do an experiment using a new technology called spectroscopy to test barley and determine varieties. We essentially wanted to build an algorithm based on regression analysis using a new testing method. The result would create a new tool for malting companies to use for testing.

One of my employees, Miranda Carr, worked with a spectroscopy unit called LabFlow from Stream Technologies. We've obtained seven hundred samples of barley from a local maltster, or barley purchaser, who has been using traditional methods to ascertain the protein levels.

Their samples have known protein levels at specific increments of one-tenth of a percent of protein. We ran those samples three times each through the LabFlow spectroscopy unit to create an algorithm that's *teaching* the machine what barley looks like at 11.8 percent protein, 12.5 percent, 13 percent, and so on.

Running twenty-one hundred samples through a spectroscopy unit is just the beginning. Once we get enough sample data into the machine and start using it on a consistent large-scale basis, the machine will correct itself over time to the point where that spectroscopy unit will be far more accurate than any other analytical method available.

First we had known data, then, using regression analysis, we were able to build an algorithm. Over time the machine itself would "learn" and by tossing out outlier data, the accuracy of the data would climb.

At the time of writing, we have about an 80 percent accuracy on protein and a 97 percent accuracy on the machine picking out different barley varieties. We began having known quantities of something in a sample, and with regression analysis we build algorithms that could turn into machine learning. This is machine learning in action.

READY FOR CHANGE?

Everyone is excited about progress, but no one wants to change. Change is hard, even if you know it is inevitable and in your best interest.

Today, the most important decision a farmer will make, in the short term, is which data platform to adopt. Have you ever been really excited when you bought something only to realize that getting it to work was going to be hard work? Farmers are no different and this is especially true of a data system.

Farmers will need to embrace data systems to stay in business. I would argue that if a farm doesn't make a migration to a data platform and develop their ability to utilize it deeply in their operation, that they won't be around in ten years. The ability to make decisions from data far surpasses the decisions we make based on our gut.

As farmers embrace the data systems, they will be able

to provide continuity of data from the seeding all the way through harvest. That data can provide information to the consumer about how the product was grown. The data can provide proof that the food was grown organically or better yet, sustainably.

Those data sets could provide substantiation to develop a sustainability index. Ultimately, I'd like to see consumers look for a sustainability metric or label on the package, more than organic or non-GMO.

TRANSPARENCY AND SUSTAINABILITY

What is transparency? Say you are traveling away on a business trip. The expectation is when you wake up, you call your spouse to let them know what you are doing for breakfast, you text who you are meeting for lunch, and you check in to mention who you are going to dinner with. You are being very transparent. But chances are you are doing all this because your spouse doesn't trust you.

ROB'S RANT

That's the thing about trust. The more you trust, the less data you need. Today's consumers are so disconnected from the farm that they don't understand the system, therefore they want lots of data. And data can be a good thing both from a transparency standpoint, particularly regarding food safety and, more importantly from a sustainability standpoint.

At the farm level, connecting the data points through a data platform will allow us to start developing the metrics around a sustainability index. I think this is something that you as a consumer can get excited about.

The farmer can begin to answer questions such as:

- How financially viable is the farm?
- What parts of the farm or the field make money; what parts do not?
- What is the utilization of fuel on a per-acre basis?
- How does that translate into greenhouse gases?
- How many times have we passed over the land?
- Can we do so in a way that reduces soil compaction?
- Can we improve the condition of the soil so that it is increasing in organic matter?

To me, sustainability is about the long-term viability of the farming system. Will it stand the test of time? Is it a system that will minimize the impact on the environment or even improve the environment?

Transparency is about proving it.

I would argue that consumers really don't know what they want. Hopefully, after reading this book you will have a clearer understanding of the factors that matter. Soil degradation, water use efficiency, nitrogen use efficiency, total pesticide load; farm viability; these are important things.

With transparency and sustainability data, would you, as a consumer, make better decisions? Let's see.

Take non-GMO cotton versus GMO cotton—which is more sustainable? Which is better for the environment?

The pesticide load on non-GMO cotton can be as high as twenty-four pounds of active pesticide per acre, versus four pounds of active pesticide per acre for GMO cotton because the GMO crop is herbicide tolerant, meaning only one or two sprays of Roundup herbicide (significantly reducing tillage), and if the cotton is also Bt, the farmer does not have to spray insecticide repeatedly on the crop to control insects attacking the cotton boll.

Which cotton is more sustainable? Who is being more transparent?

What about organic non-GMO cotton? Well, how much tillage is involved to control weeds? What is that doing to the soil? Is that sustainable? And what are the organic guys using to control bugs? What is the load of organic pesticide being used to control pests on the organic cotton? How does that stack up?

And who is being transparent? And who is truly sustainable?

The consumer today has a desire to know how the crop is grown, therefore we need to collect a large amount of information on the farm. It's incumbent upon the farmer to develop systems that can collect the data to provide the consumer with the information they want: food attributes, sustainability index tracking, and production information. The data allows the consumer to connect the dots from the loaf of bread back to how the wheat was grown in the field, to how the cotton was grown for the T-shirt.

Connecting all of the peripheral sensory devices together with data flow is part of providing the information that the consumer's trust depends on and is a big part of what we'll see in the future.

To me, the sustainability index is far more important than the non-GMO organic sticker on a T-shirt.

All the label tells you is that you're going to pay an inflated price for the T-shirt to justify the stickers, but it doesn't tell you whether the farm will be in business by the time the T-shirt fades from black to grey.

THE HOLY GRAIL

We have agreed that farm viability is important. Consumers want what they want, and this is constantly changing.

To meet these needs, we are going to require three data streams to work in concert. This is not simple. We must have:

1. Farm management data that will manage all operations on the farm
2. Transparency or traceability data that flows from the farm to the consumer
3. Sustainability index data that can be connected to the supply chain

This will require collaboration. Walmart cannot give the consumer the information they are looking for without the data flowing from the farm through the distribution channel to packager to the store.

Cracking this wicked problem will require multiple stakeholders working together. This will be tough as agriculture has traditionally been a siloed sector where a farmer is reluctant to share information with a contractor because there might be a loss of bargaining position. This scenario is replicated up the channel.

If we are going to capture more value for the farmer, it will be by providing customized solutions (like functional foods) for consumers. This will mean we will have to prove what we say on the label is true.

Organizations such as IBM Food Trust are endeavoring to connect the food chain using a "blockchain" (a computerized ledger system) that solidifies the data attached to food as it is handed off through the value chain.

Others, such as Provisional Analytics, argue that blockchain is far too linear, and that a more graphic tracking system would be better, especially when a grain, like corn, is fractured into constituents that end up in multiple food streams.

As we saw in the opening of the chapter, IBM has entered the agriculture space. Microsoft, Google, Telus, Verizon, and many other nontraditional organizations are poised to play in the agriculture space, too. This is good news, because it is through fresh lenses that we will see solutions previously unnoticed.

It will be through collaboration and openness to truly

tackle the issue of 2050 that we will ensure food security and sustainability for all.

CONCLUSION

FUTURE CONSIDERATIONS

I'm not alone in thinking about sustainability. On May 2, 2017, together with a small group of global agriculture entrepreneurs, I had the privilege of spending six hours with Bill Gates, discussing the future of agriculture.

The discussion focused on how to utilize a wide array of technologies to make a difference for the smallest landholders on the planet. We talked about how to use these ideas to lift small landholders from a state of poverty to a state of prosperity while ensuring greater sustainability.

By learning from other farming areas, could these farmers leap-frog from subsistence to prosperity? Would it be possible for these farmers to bypass the painful stages of agriculture just as Kenya bypassed copper telephone lines and the credit card system whereby everyone uses their cell phone for transactions and communication?

Bill Gates was thinking about how data systems, sensors, and connectivity can make a difference to small landholding farmers—that's why he invited us to the table to have the discussion. That session did sow some seeds indeed.

The irony is, whether you are a large or small farmer, the more information that's available on the internet, the more likely it is to confuse rather than to address your specific question.

Farmers need quality, correct information, not someone's opinion on YouTube. I still believe that at the end of the day, farmers still want to talk to a subject matter expert before they pull the trigger on a major decision.

As we look to feed the future, what are the assumptions that will impact farmers?

FUTURE ASSUMPTIONS

Here is what I see in the crystal ball for the farm of the future:

1. The next thirty years are the most critical in agriculture's history.
2. Due to economies of scale, farms will continue to get bigger.
3. Experts that have deep agriculture domain knowledge are decreasing.
4. Information grows, but solutions to difficult problems are harder to obtain.
5. We are assuming ubiquitous rural cellular/broadband coverage will come.
6. As "high-tech" sensor/data increases, farmers will still seek a "high-touch" gut check.
7. External society will place more pressure on farmers' decisions regarding food production.

When working with farmers, this is the lens through which I see the world. If asked to help guide a farm's strategic plan, these assumptions are foundational for the farm of the future.

THE MOONSHOT CHALLENGE

Like IBM, I've been thinking about how to get better information to farmers. This next section is about a new idea I have. I hope you are OK with me sharing it with you.

That 2017 session with Bill Gates, together with my time visiting Nigerian farms with Chief Audu, now Nigeria's Minister of Agriculture, sparked something in me.

It got me thinking that perhaps the scarcest resource for a farmer is not water, or nitrogen, or even money. Maybe the scarcest resource for a farmer is access to knowledge and wisdom when he/she needs it? It may be that knowledge and wisdom provided to farmers in a timely manner may be the key to nourishing the planet and ensuring infinite sustainability.

At Abundance 360, Peter Diamandis challenges us to think about "moonshots." He has defined a moonshot as something that will affect a billion people. Dan Sullivan at Strategic Coach refers to this as "game changer" thinking. Both ask the question, "How can you leverage what you know to change the world for the better by influencing a billion people?"

Since selling Agri-Trend, I have been giving this a lot of thought; is there a way for a farmer to connect with experts to get answers now?

ANSWERS NOW! AGVISORPRO

In 2018, I founded a new company called AGvisorPRO Inc. AGvisorPRO is a communication platform that aims to instantaneously connect farmers anywhere on the planet with experts to give them answers now. AGvisorPRO is the uberization of agricultural knowledge and wisdom.

Under the leadership of teammate Patrick Walther, we have been working with programmers in Calgary, Seattle, and Vietnam to build a system enabling agriculture experts (AGvisors), to stretch their brains, not their bodies.

The farmer enters his profile, the crops he grows, his livestock, equipment, practices, etc., then an algorithm would provide him instantaneous connectivity to an expert whose experience profile matches the issue at hand. These experts could include PhD-level scientists, deep domain experts, professional agronomists or technicians, mental health professionals, industry personnel, government, and academics, as well as farmers who can help other farmers.

AGvisorPRO does three things; it matches (based on algorithms), it instantaneously connects on any iOS, Android, or desktop device using ORTC (Optical Real Time Connectivity) technology and it transacts, providing clearance of payment for advice rendered. Just like Uber, it is an on-demand service where users pick their ride based on the level of expertise they are seeking and afterward, both parties can rate each other based on the interaction.

AGvisorPRO connects the shrinking resources of people who really understand agriculture to farmers who are looking for quality agricultural advice—I hope you check it out: http://www.agvisorpro.com/.

BRINGING IT ALL TOGETHER

As a consumer, at the end of the day you ultimately care about what this means to you.

Here are some considerations:

- In the short term, you want food to satisfy what you personally require during your lifespan on Earth. From a selfish perspective, you want whatever food you put in your body to serve your body. You're also concerned about your family; you want your food to be safe—you want it to taste good and you want it for a reasonable price.
- In the longer term, you might also be concerned about the infinite sustainability of agriculture. You should care about the policies and issues that will shape agriculture in the future. If today's policies are wrong, we're headed for heartbreak in the future.

How do we address both the short- and long-term objectives?

In 2008, a question was posed to a gathering of scientists of the Copenhagen Consensus. The question:

"If you were to address the key issues facing humanity and you had billions to spend, what issues would you tackle?"

Here are the results in order of priority:

1. Micronutrient supplements for children (vitamin A and zinc)
2. The Doha development agenda
3. Micronutrient fortification (iron and salt iodization)
4. Expanded immunization coverage for children
5. Biofortification

Iron deficiency, zinc deficiency, vitamin A deficiency, and iodine deficiency—these are the BIG things. If agriculture is going to be a part of the solution, we should be asking ourselves HOW are we going to employ science to address these challenges. If we need zinc or iron in our food, is there a way to add fertilizer or genetically engineer crops so that they will have higher levels of nutrition? Or will we let dogma and ideology rule the day?

INFINITE SUSTAINABILITY

I am glad you picked up this book. I believe the issues raised are important for the long-term health of agriculture and humanity. I hope that you have learned something about broad-acre farming; that you understand the 0.1 to 0.2 percent of the population working on farms in remote rural parts of North America.

We need you to understand them better because it is people like you that will influence policy that affects the livelihoods of these farm families making a living in agriculture.

We're entering the most critical phase of agriculture on the planet today, and we should be talking about the possibilities for farming and food.

- How can we help emerging farmers to leap forward to prosperous agriculture?
- How do we ensure that soil is healthy and rich?
- What are ways to match the foods we eat to our body's specific requirements?
- How do we create better value for consumers?
- How do we make it more sustainable?

ROB'S RANT

I am concerned that if today, you were to put GMO, fertilizer, pesticides, and even robotics to yes or no vote, all would be banned. People would be voting out of ignorance and fear. That vote would affect every farmer who works to raise the food we need to sustain the planet.

Next time you are out for dinner and the conversation turns to food religion, I hope you are able to put up your hands and ask, "Maybe a better issue to discuss is how do we make agriculture infinitely sustainable."

In closing, let me say, I profoundly believe in the abilities of farmers to supply you with a safe, reliable, affordable, environmentally sustainable food supply. Will you let them?

THANK YOU!

Thank you for buying and reading this book; it really means a lot that you would allow me to share my thoughts with you.

It is my hope that Food 5.0 brought you hope for the future.

I have immense faith in our farmers to feed the future. We just have to let them.

CONTACT FOR KEYNOTES OR CONSULTING

If I can help you get the message of farming out to a broader audience, such as using my services as a keynote speaker or strategic business facilitator, please let me know.

Rob@RobertSAIK.com c) +1-403-391-0772

Executive Assistant, (Shelley): SMyers.EA@gmail.com

ABOUT THE AUTHOR

ROBERT SAIK, a distinguished agrologist and professional agriculture consultant is an agricultural entrepreneur and thought leader.

As CEO of Saik Management Group Inc., he provides consulting services to many organizations and serves on several boards and advisory committees as well as facilitating The PowerFARM peer groups for progressive farmers.

Robert is CEO of Dot Farm Solutions, which is the marketing arm for the Dot autonomous robotic farming platform. In this capacity, Robert is leading the transformation of broad-acre agriculture by creating a new retail system bringing robotic technology, data, and advisory services to farmers.

He is also the founder of AGvisorPRO, a knowledge

platform providing instantaneous connectivity between farmers and agricultural experts providing answers to questions now! AGvisorPRO is open to all those with expertise in agriculture.

He serves as a technical industry advisor to Olds College, providing strategic leadership in the development of the Olds Smart Farm Ecosystem.

Saik is a partner in a farming operation in Uganda and he has pure-bred cattle in Saskatchewan.

Robert is a partner in Perigro Inc., a venture capital firm specializing in investment in agriculture and technology companies.

He is a member of the A100 (Alberta Tech Entrepreneur Network), a student of Dan Sullivan, The Strategic Coach and a member of Abundance 360 with Peter Diamandis.

INFORMATION

WEBSITES: TWITTER:

www.RobertSAIK.com   @RSaik
www.seeDOTrun.com   @seeDOTrun.com
www.AGvisorPRO.com   @AGvisorPRO
www.linkedin.com/in/robertsaik/

TEDx—Will Agriculture be ALLOWED to FEED 9 Billion People? https://www.youtube.com/ watch?v=xvFD6DRnoCg

BACKGROUND

Robert's technical strengths lie in soil chemistry, plant physiology, and crop nutrition.

He has published over fifty articles on crop agronomics and is a thought leader on the integration of technology in crop production.

In 2017, Robert, as part of an international delegation, was invited to present his ideas to Bill Gates on how agriculture technology could play a role in lifting small landholders out of poverty and into prosperity.

He founded Agri-Trend in 1997, a professional agricultural coaching network comprised of over 150 consultants backed by over thirty PhDs and Masters of Science. Simultaneously the Agri-Data Solution was created to enable the online management of agronomic, precision agriculture, grain marketing, farm business management, and carbon credit data for use by farmers and consultants. In 2015, the Agri-Trend group of companies was acquired by Trimble Inc. Robert provided global business development leadership for Trimble until 2018.

Rob has played a foundational role in the introduction of many new technologies into the agricultural sector ranging from the use of elemental sulfur for soil amelioration to the use of copper in wheat, boron in canola, use of satellite imagery for variable-rate fungicide application to the development of full mobile data systems on the farm.

In 2001, Saik wrote his first position paper on the role reduced tillage can play on carbon sequestration. He played a key consultative role in the development of Alberta's Carbon Offset Protocols and in 2007 founded Agri-Trend Aggregation, an agriculture carbon credit company. This ISO certified organization works in the area of tillage, beef, and nitrous oxide emission reduction carbon credit offset creation.

Agri-Trend was recognized as a 2012 and 2014 Western Regional Canada Top 50 Best Managed Company and was recognized by Venture Magazine as one of Alberta's 2013 Top 25 Most Innovative Organizations.

Robert has created many educational processes to provide training for agricultural professionals. His sessions have been attended by thousands. His systems have been implemented on millions of acres by hundreds of agronomists. He has delivered agronomic and technical talks in Ukraine, Australia, Argentina, United Arab Emirates, the UK, the EU, Mexico, and throughout North America.

Robert is also the author of an Amazon 2014 Best of Books, *The Agriculture Manifesto—Ten Key Drivers That Will Shape Agriculture in the Next Decade*. His was also the founder and publisher of The AgADVANCE—Journal for Growing Innovations, 2014 Western Trade Journal of the Year.

He is the creator of the Agri-Prize contest series, a skills-based incentive competition for agriculture and launched the Canola 100 Agri-Prize in July 2015, which is a collaboration with Agri-Trend, John Deere, and Glacier FarmMedia with the goal of attaining a yield of 100 bu/ac of spring sown, dryland canola on fifty contiguous acres.

He is a passionate keynote speaker addressing audiences on the importance of modern agriculture. His 2014 TEDx talk, entitled "Will Agriculture be ALLOWED to feed 9 Billion People" is a popular online reference and has been viewed over 140,000 times. His other keynotes include "GMO, what you know ain't so," "The Agriculture Manifesto," "Food 5.0—Convergence on the Farm" and how to manage your business through the "Winds of Change."

He is the executive producer of KNOW IDEAS Media (Facebook/YouTube) which is dedicated to providing science-based education on the use of genetic engineering in agriculture. This project produces multiple initiatives including a webisode series and vignettes on GMO technology for use by teachers at www.know-GMO.ca.

Robert is a passionate entrepreneur and has founded over fifteen companies ranging from field diagnostic and sensory technology firms, fertilizer manufacturing, fertilizer distribution, agri-retail, agricultural consulting, and farming, including a purebred cattle operation.

Internationally Robert has been involved in projects across North America, Mexico, United Arab Emirates, the UE, Ukraine, Kazakhstan, Australia, Argentina, Kenya, Nigeria, and Uganda.

Through his work with A Better World, Rob has been involved in projects bringing irrigation to orphanages and he has conducted a soil inventory in both Kenya and Uganda to identify nutrient deficiencies in soils.

He served as a member of the Telus Agriculture Advisory Council providing input on how technology can be leveraged to improve agriculture productivity.

Most recently, Robert has been appointed by the Province of Alberta to sit on the Talent Advisory Council on Technology (TACT). TACT will focus on the strategic development implementation plans directed to increase technology training, adaptation, and systems integration in Alberta's universities, colleges, technical and corporate training curriculums.

Rob serves on several boards and is a member of many professional associations. In 2006, he was recognized as Distinguished Agrologist of the Year by the Alberta Institute of Agrology.

Robert was awarded the 2014 Canadian Agri-Marketer of the Year by the Canadian Association of Agricultural Marketers.

Robert resides in Olds, Alberta, Canada.